The ASVAB Tutor's Mathematics Knowledge Study Guide
by
Julie A. Hyers

Julie A. Hyers © 2020

The ASVAB Tutor's Mathematics Knowledge Study Guide
Julie A. Hyers © 2020

The Mathematics Knowledge section of the ASVAB consists of various topics in mathematics.

The book has broken the section into 15 subtopics and consists of:

- Mathematics formulas to know
- 15 introductory lessons covering each subtopic
- 15 pre-tests with explanations of answers
- 15 videos solving the pre-test questions (online access)
- 15 post-tests with explanations of answers

15 Mathematics Knowledge Subtopics:
1. Signed Numbers and Rounding
2. Squares and Cubes
3. Prime and Composite Numbers and Average
4. Triangles
5. Angles and Slope
6. Area, Circumference, and Perimeter
7. Volume and Surface Area
8. Real Life Applications of Perimeter, Area, Surface Area, and Volume
9. PEMDAS - Order of Operations, Scientific Notation, and Factorial
10. Exponents
11. Factoring and FOIL (Distribution) in Algebra
12. Solve for x
13. Basic Operations with Algebra
14. Algebra with Fractions
15. Celsius, Fahrenheit, and Absolute Value

*Not responsible for typographical errors.

Julie A. Hyers © 2020

Table of Contents

Mathematics Knowledge Review, Pre-Tests and Answer Keys pages 4-119

Mathematics Formulas to Know for Test pages 4-5

1. Signed Numbers and Rounding
   - Review, Pre-Test, Answer Key pages 6-12
2. Squares and Cubes
   - Review, Pre-Test, Answer Key pages 13-19
3. Prime and Composite Numbers and Average
   - Review, Pre-Test, Answer Key pages 20-27
4. Triangles
   - Review, Pre-Test, Answer Key pages 28-32
5. Angles and Slope
   - Review, Pre-Test, Answer Key pages 33-43
6. Area, Circumference, and Perimeter
   - Review, Pre-Test, Answer Key pages 44-55
7. Volume and Surface Area
   - Review, Pre-Test, Answer Key pages 56-62
8. Real Life Applications of Perimeter, Area, Surface Area, and Volume
   - Review, Pre-Test, Answer Key pages 63-70
9. PEMDAS - Order of Operations, Scientific Notation, and Factorial
   - Review, Pre-Test, Answer Key pages 71-76
10. Exponents
    - Review, Pre-Test, Answer Key pages 77-82
11. Factoring and FOIL (Distribution) in Algebra
    - Review, Pre-Test, Answer Key pages 83-90
12. Solve for x
    - Review, Pre-Test, Answer Key pages 91-101
13. Basic Operations with Algebra
    - Review, Pre-Test, Answer Key pages 102-106
14. Algebra with Fractions
    - Review, Pre-Test, Answer Key pages 107-113
15. Celsius, Fahrenheit, and Absolute Value
    - Review, Pre-Test, Answer Key pages 114-119

Mathematics Knowledge Post-Tests pages 120-186

1. Signed Numbers and Rounding
    - Post-Test, Answer Key pages 121-124
2. Squares and Cubes
    - Post-Test, Answer Key pages 125-128
3. Prime and Composite Numbers and Average
    - Post-Test, Answer Key pages 129-133
4. Triangles
    - Post-Test, Answer Key pages 134-136
5. Angles and Slope
    - Post-Test, Answer Key pages 137-143
6. Area, Circumference, and Perimeter
    - Post-Test, Answer Key pages 144-149
7. Volume and Surface Area
    - Post-Test, Answer Key pages 150-152
8. Real Life Applications of Perimeter, Area, Surface Area, and Volume
    - Post-Test, Answer Key pages 153-157
9. PEMDAS - Order of Operations, Scientific Notation, and Factorial
    - Post-Test, Answer Key pages 158-161
10. Exponents
    - Post-Test, Answer Key pages 162-165
11. Factoring and FOIL (Distribution) in Algebra
    - Post-Test, Answer Key pages 166-170
12. Solve for x
    - Post-Test, Answer Key pages 171-176
13. Basic Operations with Algebra
    - Post-Test, Answer Key pages 177-179
14. Algebra with Fractions
    - Post-Test, Answer Key pages 180-183
15. Celsius, Fahrenheit, and Absolute Value
    - Post-Test, Answer Key pages 184-186

Mathematics Knowledge Reviews, Pre-Tests, and Answer Keys

Mathematics Knowledge
Formulas To Know

Shapes

Name of Shape	Number of Sides	Name of Shape	Number of Sides
triangle	3	octagon	8
quadrilateral	4	nonagon	9
pentagon	5	decagon	10
hexagon	6	hendecagon/undecagon	11
septagon/heptagon	7	dodecagon	12

Perimeter
To find the perimeter of any shape, add up the lengths of all the sides.

Area Formulas
Area of a square = s^2 = side x side

Area of a rectangle = lw = length x width

Area of a triangle = 1/2 bh = 1/2 x base x height

Area of a parallelogram = bh = base x height

Area of a trapezoid = 1/2 $(b_1 + b_2)$h = 1/2[(base 1 + base 2) x height]

Circle Formulas:
Circumference   C = πd = pi x diameter = 3.14 x diameter
Area of a circle   A = $πr^2$ = pi x radius squared = 3.14 x radius x radius

Diameter is the distance across the center of the circle.
Radius is the distance from the center of the circle to the outside of the circle.
Diameter is twice the length of the radius.
π = pi = 3.14

Volume Formulas
Volume of a cube $V = s^3$ = side x side x side

Volume of a rectangular box $V = lwh$ = length x width x height

Volume of a cylinder $V = \pi r^2 h$ = pi x radius x radius x height

Volume of a sphere (ball) $V = 4/3\, \pi r^3$ = 4/3 x pi x radius x radius x radius

Simpler sphere formula is an estimate of the real formula
$V = 4r^3$ = 4 x radius x radius x radius

The real answer will be a bit bigger than the simpler sphere formula.
Only use this if the answer choices on the test are not close.
If the answer choices have values that are near to each other, do not use this estimate.
The simpler sphere formula gives an estimate of the real answer.
Pick the choice that is a little bigger. This simpler formula can save a lot of time on the test.

Surface Area Formulas
Surface area of a cube = $6s^2$ = 6 x side x side

Surface area of rectangular box = 2(LW + LH + WH) =
2[(length x width) + (length x height) + (width x height)]

Pythagorean Theorem
This formula is used to find the measurement of a side of a right triangle when the other 2 sides are given.
$c^2 = a^2 + b^2$

Common Measurements of Right Triangles

Knowing the measurement of common right triangles can help you to avoid using Pythagorean theorem at times.

3:4:5   5:12:13
6:8:10   10:24:26

Slope Formula
slope = $\dfrac{y_2 - y_1}{x_2 - x_1}$ = $\dfrac{\text{change in y}}{\text{change in x}}$

Julie A. Hyers © 2020

Signed Numbers and Rounding Review

Basic Operations with Positive and Negative Numbers

Adding Positive and Negative Numbers

When adding numbers with the same signs, add the numbers and keep the sign.

Positive + Positive = Positive
Example: 6 + 4 = 10

Negative + Negative = Negative
Example: -6 + -4 = -10

When adding numbers with different signs, subtract the numbers and keep the sign of the bigger number.
Example: -4 + 6 = 2
or -6 + 4 = -2

Subtraction of Positive and Negative Numbers

Subtracting a positive number from a positive number:
This is normal subtraction.
Example: 6 – 4 = 2

Subtracting a negative number from a negative number:
When you see a subtraction sign followed by a negative sign, the two negative signs become a plus sign. Subtract the 2 numbers and take the sign of the bigger number.
Example: -6 - (-4) =
-6 + 4 = -2

Subtracting a positive number from a negative number:
Since there is a negative in front of the first number and a subtraction in front of the second number, add the 2 numbers and keep the negative sign.
Example: -4 - 6 = -10

Subtracting a negative number from a positive number.
Turn the 2 negatives into 2 positives. It becomes an addition problem.
Example: 6 – (-4) = 6 + 4 = 10

Multiplication of Positive and Negative Numbers

When the signs are the same, the answer is positive.

Positive x Positive = Positive
Example: 7 x 8 = 56

Negative x Negative = Positive
Example: -7 x -8 = 56

When the signs are different, the answer is negative.

Positive x Negative = Negative
Example: 7 x -8 = -56

Negative x Positive = Negative
Example: -7 x 8 = -56

Division of Positive and Negative Numbers

When the signs are the same, the answer is positive.

Positive ÷ Positive = Positive
Example: 42 ÷ 6 = 7

Negative ÷ Negative = Positive
Example: -42 ÷ -7 = 6

When the signs are different, the answer is negative.

Positive ÷ Negative = Negative
Example: 42 ÷ -7 = -6

Negative ÷ Positive = Negative
Example: -42 ÷ 7 = -6

Place Value

1,	0	0	0,	0	0	0,	0	0	0.	.	0	0	0	0
billions	hundred millions	ten millions	millions	hundred thousands	ten thousands	thousands	hundreds	tens	ones		tenths	hundredths	thousandths	ten thousandths

To work on rounding, you must know place value.

Rounding- When rounding, circle the number in the place that it needs to be rounded to.
Then look to the number on the right.
If the number to the right of that number is 4 or below, keep the number the same.
If the number to the right of that number is 5 or above, round the number up.

Remember: 4 and below, let it go.
           5 and above, give it a shove.

Round to the nearest hundred.
     3,447.932

     3,400

Round to the nearest thousandth.
     989.1946

     989.195

Mathematics Knowledge
Signed Numbers & Rounding - Pre-Test

1. $-2 + 4 =$ 2

    $-2, -1, 0, 1, \boxed{2}$

2. $-9 + -4 =$ -13

    $-9 + -4$

3. $-8 - 8 =$ -16

4. $3 - (-7) =$ -21

    $= 10$

5. $-5 \times -6 =$ $+30$

6. $-1 \times 5 =$ -5

7, 14, 21, 28, 35, 42, 49, 56

7. $56 \div -7 =$ -8

(+) (-) = (-)

8. $-24 \div -3 =$ +8

9. Round to the nearest hundred.
2,887.876

2,900.000

10. Round to the nearest thousandth.
542.3671

542.367

Mathematics Knowledge
Signed Numbers & Rounding - Pre-Test with Answers

1. $-2 + 4 = 2$ ✓

2. $-9 + -4 = -13$ ✓

3. $-8 - 8 = -16$ ✓

4. $3 - (-7) = 3 + + 7 = 10$ ✗

5. $-5 \times -6 = 30$ ✓

6. $-1 \times 5 = -5$ ✓

7. $56 \div -7 = -8$ ✓

8. $-24 \div -3 = 8$ ✓

9. Round to the nearest hundred.
    2,887.876

    2,900 ✓

10. Round to the nearest thousandth.
    542.3671

    542.367 ✓

Squares and Cubes Review

Squares of Positive Numbers
$1^2 = 1$
$2^2 = 4$
$3^2 = 9$
$4^2 = 16$
$5^2 = 25$
$6^2 = 36$
$7^2 = 49$
$8^2 = 64$
$9^2 = 81$
$10^2 = 100$
$11^2 = 121$
$12^2 = 144$

Squares of Negative Numbers
$(-1)^2 = 1$
$(-2)^2 = 4$
$(-3)^2 = 9$
$(-4)^2 = 16$
$(-5)^2 = 25$
$(-6)^2 = 36$
$(-7)^2 = 49$
$(-8)^2 = 64$
$(-9)^2 = 81$
$(-10)^2 = 100$
$(-11)^2 = 121$
$(-12)^2 = 144$

Square Roots
$\sqrt{1} = 1$ or -1
$\sqrt{4} = 2$ or -2
$\sqrt{9} = 3$ or -3
$\sqrt{16} = 4$ or -4
$\sqrt{25} = 5$ or -5
$\sqrt{36} = 6$ or -6
$\sqrt{49} = 7$ or -7
$\sqrt{64} = 8$ or -8
$\sqrt{81} = 9$ or -9
$\sqrt{100} = 10$ or -10
$\sqrt{121} = 11$ or -11
$\sqrt{144} = 12$ or -12

Cubes of Positive Numbers
$1^3 = 1$
$2^3 = 8$
$3^3 = 27$
$4^3 = 64$
$5^3 = 125$
$6^3 = 216$

Cubes of Negative Numbers
$(-1)^3 = -1$
$(-2)^3 = -8$
$(-3)^3 = -27$
$(-4)^3 = -64$
$(-5)^3 = -125$
$(-6)^3 = -216$

Cube Roots of Positive Numbers
$\sqrt[3]{1} = 1$
$\sqrt[3]{8} = 2$
$\sqrt[3]{27} = 3$
$\sqrt[3]{64} = 4$
$\sqrt[3]{125} = 5$
$\sqrt[3]{216} = 6$

Cube Roots of Negative Numbers
$\sqrt[3]{-1} = -1$
$\sqrt[3]{-8} = -2$
$\sqrt[3]{-27} = -3$
$\sqrt[3]{-64} = -4$
$\sqrt[3]{-125} = -5$
$\sqrt[3]{-216} = -6$

Negative Signs and Exponents

Placement of the negative sign when dealing with exponents affects the answer.

$(-5)^2 = -5 \times -5 = 25$
This answer is positive because a negative x negative = positive.

$-(5^2) = -(5 \times 5) = -25$
This answer is negative because it asks for the multiplication of 2 positive numbers, which gives a positive answer. Then it is multiplied by a negative number. The answer is negative.

Simplifying Radicals

√108

Work to find a perfect square that goes into 108 to break it down.

√36 x √6 = 6√6

Radicals in the Denominator

$\dfrac{6\sqrt{4}}{\sqrt{5}}$

A radical cannot remain in the denominator.
To get rid of it, multiply by √5 on the top and bottom.

$\dfrac{6\sqrt{4}}{\sqrt{5}} \times \dfrac{\sqrt{5}}{\sqrt{5}} = \dfrac{6\sqrt{20}}{5} = \dfrac{6\sqrt{4}\sqrt{5}}{5} = \dfrac{6 \times 2\sqrt{5}}{5} = \dfrac{12\sqrt{5}}{5}$

Mathematics Knowledge
Squares and Cubes - Pre-Test

1. $\sqrt{196} =$ ~~14~~

2. $11^2 =$ 121

3. $(-10^2) =$ 100

4. $\sqrt[3]{-27} =$ 27

5. $4^3 =$ 64

6. $(-2)^3 =$

7. $(-3)^2 =$

8. $-(3^2) =$

9. $\sqrt{96}$

10. $\dfrac{10\sqrt{2}}{\sqrt{3}}$

Mathematics Knowledge
Squares and Cubes - Pre-Test with Answers

1. $\sqrt{196} =$
   A square root of a number is a number that when multiplied by itself, gives the number.
   The square root of 196 is +14 or -14 because 14 x 14 = 196 and -14 x -14 = 196

2. $11^2 =$
   11 x 11 = 121

3. $(-10)^2 =$
   -10 x -10 = 100

4. $\sqrt[3]{-27} = -3$
   The cube root of a number is a number that when used in multiplication three times results in that number.
   The cube root of -27 is -3 because -3 x -3 = -27

5. $4^3 =$
   4 x 4 x 4 = 64

6. $(-2)^3 =$
   -2 x -2 x -2 = -8

7. $(-3)^2 =$
   -3 x -3 = 9

8. $-(3^2) =$
   -(3 x 3) = 9

9. $\sqrt{96}$

Work to find a perfect square that goes into 96 to break it down.
$\sqrt{16} \times \sqrt{6} = 4\sqrt{6}$

10. $\dfrac{10\sqrt{2}}{\sqrt{3}}$

A radical cannot remain in the denominator. To get rid of it, multiply by $\sqrt{3}$ on the top and bottom.

$\dfrac{10\sqrt{2}}{\sqrt{3}} \times \dfrac{\sqrt{3}}{\sqrt{3}} = \dfrac{10\sqrt{6}}{3}$

Prime and Composite Numbers and Average Review

Prime number is a number that can only be divided by 1 and itself.
1 is not a prime number.
Examples: 2, 3, 5, 7, 11, 13, 17, etc.

Composite number is a number than can be divided by at least one other number aside from itself and 1.
Examples: 4, 6, 8, 9, 10, 12, 15, 16, 18, etc.

Divisibility Rules

Knowing divisibility rules is very useful in figuring out which numbers are prime or composite. It is also helpful in reducing fractions.

A number is divisible by:

2 if it is even. Even numbers end with: 0, 2, 4, 6, 8.

3 if the sum of the digits is divisible by 3.
    Example: 123, $1 + 2 + 3 = 6$, 6 is divisible by 3, therefore 123 is divisible by 3.

4 if the last 2 digits are divisible by 4.
    Example: 1024 is divisible by 4 because 24 is divisible by 4.

5 if the number ends with a 5 or a 0.

6 if the number is even and the sum of the digits are divisible by 3.
    To be divisible by 6, a number needs to follow divisibility rules for 2 and 3.

9 if the sum of the digits is divisible by 9. Example: 981, $9 + 8 + 1 = 18$
    18 is divisible by 9, therefore 981 is divisible by 9.

10 if the last digit of a number is 0.

Mean, Median, Mode

Mean is another word for average.
Add up the digits and divide by the number of digits to find the mean.
Example: 72, 84, 92, 65, 54.
72 + 84 + 92 + 65 + 54 = 367
367 ÷ 5 = 73.4
The mean or average is 73.4

Median is the middle number in a list of numbers when the numbers are listed from least to greatest.

If there is an odd amount of numbers, the median is the middle number.

Example: 24, 37, 39, 42, 51
There is an odd amount of numbers. The median is the middle number.
The median is 39.

If there is an even amount of numbers, find the average of the 2 middle numbers to find the median.

Example: 34, 45, 48, 50, 53, 60
In this case, there is an even amount of numbers. The 2 middle numbers are 48 and 50.
Find the average of 48 and 50.
48 + 50 = 98
98 ÷ 2 = 49
The median is 49.

Mode is the most common number in a list of numbers.
Example: 8, 8, 10, 12, 14, 15
The mode is 8.

Sometimes there can be more than one mode in a list of numbers.
If there are 2 modes, it is known as bimodal.
Example: 5, 6, 6, 7, 7, 8
This example has 2 modes.
The 2 modes are 6 and 7.

Prime Factorization

Prime Factorization involves breaking down a number into its factors, which are prime numbers.

Example: What is the prime factorization of 100?

This can be done a few different ways.

Since 100 has a number of different factors, start with any 2 factors.
Keep dividing until all the factors are prime numbers.
It does not matter which 2 factors you choose first.

100 = 10 x 10
100 = 2 x 5 x 2 x 5
100 = 2^2 x 5^2

If you chose different factors for 100, you would still end up with the same answer in the end.

100 = 25 x 4
100 = 5 x 5 x 2 x 2
100 = 2 x 2 x 5 x 5
100 = 2^2 x 5^2

Julie A. Hyers © 2020

Mathematics Knowledge
Prime & Composite Numbers, Average - Pre-Test

1. Which of the following is a prime number?
   A. 401
   B. 402
   C. 405
   D. 411

2. Which of the following is a composite number?
   A. 23
   B. 27 ← 3 × 9
   C. 29
   D. 31

   factor × factor = product

3. Which of the following are similar figures?
   A. lion and tiger
   B. letter and envelope
   C. building and scale model of building
   D. jar and lid

4. What is the mean of these numbers?
   10, 12, 14, 19, 19, 20

   10+12 = 22   36+14
   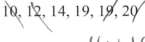
   14

5. What is the median of these numbers?
   10, 12, 14, 19, 19, 20   16.5

   14+19

23

6. What is the mode of these numbers?
   10, 12, 14, 19, 19, 20

   14

7. What is the prime factorization of 80?

   $2^4 \times 5$

8. $\dfrac{9 \text{ ft} + 7 \text{ yd}}{2} =$

9. What is the average of 1/6 and 1/8?

Julie A. Hyers © 2020

Mathematics Knowledge
Prime & Composite Numbers, Average - Pre-Test with Answers

1. Which of the following is a prime number?
   A. 401
   B. 402
   C. 405
   D. 411

   Know the divisibility rules for 2, 3 and 5.

   A number is divisible by 2 if it is even. Eliminate even numbers. The only even number that is prime is the number 2. Eliminate 402.

   A number is divisible by 3 if the sum of the digits is divisible by 3.
   411 is 4 +1 + 1 = 6 and 6 is divisible by 3, therefore 411 is divisible by 3. Eliminate 411.

   A number is divisible by 5 is the last digit is 5 or 0.
   405 ends with a 5, so it is divisible by 5. Eliminate 405.

   The remaining answer is the prime number, A. 401.

2. Which of the following is a composite number?
   A. 23
   B. 27
   C. 29
   D. 31

   A composite number is a number that can be divided by at least one other number aside from 1 and itself. 27 is divisible by 3 and 9, and therefore it is a composite number. The answer is B. 27.

3. Which of the following are similar figures?
   A. lion and tiger
   B. letter and envelope
   C. building and scale model of building
   D. jar and lid

   Similar figures are objects that have the same shape but may have the same size or different sizes.
   Choice C, the building and scale model of the building are similar figures.

4. What is the mean of these numbers?
   10, 12, 14, 19, 19, 20

   To find the mean (average), add up the numbers and divide by how many numbers there are.
   10 + 12 + 14 + 19 + 19 + 20 = 94
   94 ÷ 6 =
   Mean = 15.666

5. What is the median of these numbers?
   10, 12, 14, 19, 19, 20

   To find the median of a list of numbers, list the numbers in order from least to greatest. The middle number is the median. When there is an even amount of numbers, take the 2 middle numbers, add them up and divide by 2.
   14 + 19 = 33
   33 ÷ 2 =
   Median = 16.5

6. What is the mode of these numbers?
   10, 12, 14, 19, 19, 20

   The mode is the most common number in a list of numbers.
   In this case, the mode is 19.

7. What is the prime factorization of 80?

   To find the prime factorization of 80, divide the number up until all the factors are prime.
   80 = 40 x 2
   80 = 10 x 4 x 2
   80 = 5 x 2 x 2 x 2 x 2
   80 = 2 x 2 x 2 x 2 x 5
   The prime factorization of 80 is 2^4 x 5

8. $\dfrac{9 \text{ ft} + 7 \text{ yd}}{2} =$

   Change the yards to feet.
   There are 3 ft in a yard.
   9 ft + 21 ft = 30 ft
   30 ft/2 = 15 ft

9. What is the average of 1/6 and 1/8?

   To find the average of 1/6 and 1/8, add the two fractions and divide by 2.

   If you use the least common denominator of 24, here is the answer.
   1/6 + 1/8 =
   4/24 + 3/24 = 7/24
   7/24 ÷ 2 = 7/24 x 1/2 = 7/48

   If you use 48 as a common denominator, you end up with the same answer in the end.
   1/6 + 1/8
   8/48 + 6/48 = 14/48 = 7/24
   7/24 ÷ 2 = 7/24 x 1/2 = 7/48

Triangles Review

All the angles in a triangle add up to 180 degrees.

Triangles

Acute- a triangles with all angles less than 90 degrees. (The sum of the angles is still 180°.)

Right- a triangle with one 90-degree angle. (The sum of the angles is still 180°.)

Obtuse- a triangle with one angle greater than 90 degrees. (The sum of the angles is still 180°.)

Triangles

Equilateral Triangle - all 3 sides and all 3 angles are equal. Each angle is 60 degrees.

What is the measure of an angle in an equilateral triangle?

An equilateral triangle has 3 equal angles. The total measure of the angles in a triangle is always 180°. An equilater`1al triangle has 3 equal angles (and 3 equal sides) making each angle a 60° angle.

Isosceles Triangle - has 2 equal sides and 2 equal angles.

If one of the equal angles of an isosceles triangle is 70°, what is the measure of the vertex angle?

An isosceles triangle has 2 equal sides and 2 equal angles. If one of the equal angles is 70°, the other equal angle is 70°. To find the final angle, subtract the measure of the other 2 angles from 180°.

70 + 70 = 140° and

180° - 140° = 40°. The measure of the vertex angle is 40°.

Scalene Triangle - all 3 sides and all 3 angles are different.

Right Triangle- one angle in a triangle is 90 degrees.
In a right triangle, if one of the angles measures 40°, what is the measure of the third angle?

In a right triangle, one angle is 90°. The other angle is 40°. To find the third angle, add up 90° + 40° = 130° and subtract it from 180°. 180° - 130° = 50°.

Sides of a Right Triangle
Legs are the two shorter sides of a right triangle.
Hypotenuse is the longest side of a right triangle. It is the side opposite the right angle.

Pythagorean Theorem
Used to find the length of the sides of a right triangle.
$c^2 = a^2 + b^2$
c = hypotenuse
a and b are the legs

Use the Pythagorean theorem to solve this question about the sides of a right triangle.

If a right triangle's legs have the measures of 30 and 40, find the measure of the hypotenuse.
$c^2 = a^2 + b^2$
$c^2 = 30^2 + 40^2 = 900 + 1600 = 2500$
$c^2 = 2500$
$c = \sqrt{2500} = 50$

This is a 30:40:50 triangle.

The 30:40:50 is the 10 times the size of the 3:4:5 triangle.

To make things easier, memorize the common right triangles.

Memorize the common measurements of right triangles below.
It can help save time so you can avoid using Pythagorean theorem at times.
If you are given questions that include the measurements below or any multiple of those measurements, you can solve the problem without Pythagorean theorem.

Common Measurements for Right Triangles

3:4:5	5:12:13
6:8:10	10:24:26

Mathematics Knowledge
Triangles - Pre-Test

1. What is the measure of an angle in an equilateral triangle?

2. If one of the equal angles of an isosceles triangle is 50°, what is the measure of the vertex angle?

3. In a right triangle, if one of the angles measures 20°, what is the measure of the third angle?

4. If a right triangle has sides that measure 6 and 8, what is the measure of the longest side?

   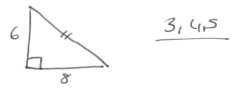

   3, 4, 5

5. Fred walks 5 blocks north and 12 blocks east. If a straight line is drawn from his starting point to his ending point, how many blocks is Fred from his starting point?

Julie A. Hyers © 2020

Mathematics Knowledge
Triangles - Pre-Test with Answers

1. What is the measure of an angle in an equilateral triangle?

   An equilateral triangle has 3 equal angles. The total measure of the angles in a triangle is always 180°. An equilateral triangle has 3 equal angles (and 3 equal sides) making each angle a 60° angle.

2. In one of the equal angles of an isosceles triangle is 50°, what is the measure of the vertex angle?

   An isosceles triangle has 2 equal sides and 2 equal angles. If one of the equal angles is 50°, the other equal angle is 50°.

   To find the final angle, subtract the measure of the other 2 angles from 180°. 50 + 50 = 100° and 180° - 100° = 80°.
   The measure of the vertex angle is 80°.

3. In a right triangle, if one of the angles measures 20°, what is the measure of the third angle?

   In a right triangle, one angle is 90°. The other angle is 20°. To find the third angle, add up 90° + 20° = 110° and subtract it from 180°.
   180° - 110° = 70°.

4. If a right triangle has sides that measure 6 and 8, what is the measure of the longest side?

   Use the Pythagorean theorem to solve this question about the sides of a right triangle.
   $c^2 = a^2 + b^2$
   $c^2 = 6^2 + 8^2 = 36 + 64 = 100$
   $c^2 = 100$
   $c = \sqrt{100} = 10$

   To make things easier, memorize the common right triangles

   3:4:5   5:12:13

   6:8:10   10:24:26

5. Fred walks 5 blocks north and 12 blocks east. If a straight line is drawn from his starting point to his ending point, how many blocks is Fred from his starting point?

Use Pythagorean theorem to solve this problem.
The longest side is c, and the other 2 sides are a and b.
$c^2 = a^2 + b^2$
$c^2 = 5^2 + 12^2 = 25 + 144 = 169$
$c^2 = 169$
$c = \sqrt{169} = 13$

To make this easier, make sure you remember the common right triangles.

3:4:5   5:12:13

6:8:10   10:24:26

Angle and Slope Review

Complementary angles add up to 90 degrees.

Subtract one angle from 90 to figure out the value of its complementary angle.

Example:
What is the complement of a 48° angle?
Complementary angles add up to 90°.
90° – 48° = 42°

Supplementary angles add up to 180 degrees.

Subtract one angle from 180 to figure out the value of its supplementary angle.

Example:
What is the supplement of an 85° angle?
Supplementary angles add up to 180°.
180° - 85° = 95°

Slope Formula

Formula for slope = $\dfrac{y_2 - y_1}{x_2 - x_1}$ =

Example:
If a line has the following 2 points (2,5) and (4, -1), find the slope of the line.

Formula for slope = $\dfrac{y_2 - y_1}{x_2 - x_1}$ $\dfrac{5 - -1}{2 - 4}$ = $\dfrac{6}{-2} = -3$

Slope of a Line

What is the slope of a line if the equation for the line is y = 2x + 10?

This line is in the slope intercept form y = mx + b.

m is the slope, therefore the slope is 2.

slope = 2

y-intercept of a line

What is the y-intercept of a line if the equation for the line is $y = 5x - 7$?

This line is in the slope-intercept form $y = mx + b$

b is the y-intercept (where the line crosses the y-axis)

The y-intercept is -7.

Angle Measures Find the measure of angle 7 if angle 1 = 62°. Angle 7 is _____ degrees.

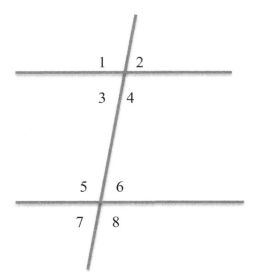

Angles that are next to each other add up to 180°.

Vertical angles, angles that are diagonally across from each other are equal.

The angle measures of angles 1, 2, 3, and 4 are repeated in angles 5, 6, 7, and 8 respectively.

If angle 1 is 62°, angle 2 is 118°,

angle 3 is 118°, angle 4 is 62°,

angle 5 is 62°, angle 6 is 118°,

angle 7 is 118°, angle 8 is 62°,

and angle 7 is 118°.

Angles

Acute angles are less than 90 degrees.

Right angles equal 90 degrees.

Obtuse angles are greater than 90 degrees and less than 180 degrees.

An Acute Angle is less than 90 degrees.

Which of the following is an acute angle?

A. 35°
B. 90°
C. 135°
D. 180°

An acute angle is less than 90°.

The answer is A. 35°

An Obtuse Angle is More than 90° and Less than 180°.

Which of the following is an obtuse angle?

A. 35°

B. 90°
C. 135°
D. 180°

An obtuse angle is bigger than 90° and less than 180°.

The answer is C. 135°

A Right Angle is Equal to 90°

Which of the following is a right angle?

A. 35°
B. 90°
C. 135°
D. 180°

A right angle is equal to 90°.

The answer is B. 90°.

Mathematics Knowledge
Angles and Slope - Pre-Test

1. What is the complement of an 82° angle?

2. What is the supplement of a 70° angle?

3. If a line has the following 2 points (0,1) and (5, -4), find the slope of the line.

4. What is the slope of a line if the equation for the line is $y = 5x + 4$?

5. What is the y-intercept of a line if the equation for the line is $y = 2x - 6$?

6. Find the measure of angle 7 if angle 1 = 105°.   Angle 7 is _____ degrees.

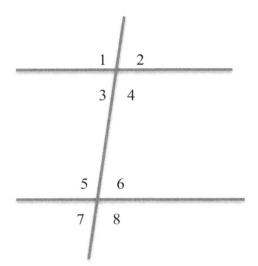

7. Which of the following is an acute angle?
   A. 65°
   B. 90°
   C. 125°
   D. 180°

8. Which of the following is an obtuse angle?
   A. 65°
   B. 90°
   C. 125°
   D. 180°

9. Which of the following is a right angle?
   A. 65°
   B. 90°
   C. 125°
   D. 180°

Mathematics Knowledge
Angles and Slope - Pre-Test with Answers

1. What is the complement of an 82° angle?

    90° − 82° = 8°

2. What is the supplement of a 70° angle?

    180° − 70° = 110°

3. If a line has the following 2 points (0,1) and (5, -4), find the slope of the line.

    slope = $\frac{y_2 - y_1}{x_2 - x_1}$ = $\frac{1 - -4}{0 - 5}$ = $\frac{5}{-5}$ = -1

    slope = -1

4. What is the slope of a line if the equation for the line is y = 5x + 4?

    The slope of the line is 5.

5. What is the y-intercept of a line if the equation for the line is y = 2x − 6?

    The y-intercept of the line is -6.

6. Find the measure of angle 7 if angle 1 = 105°.

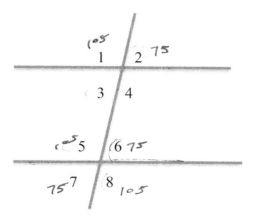

Angles that are next to each other add up to 180°.

Vertical angles, angles that are diagonally across from each other are equal.

The angle measure of angles 1, 2, 3, and 4 are repeated in angles 5, 6, 7, and 8 respectively.
If angle 1 is 105°, angle 2 is 75°,
angle 3 is 75°, angle 4 is 105°,
angle 5 is 105°, angle 6 is 75°,
angle 7 is 75°, and angle 8 is 105°.

Angle 7 is 75°.

7. Which of the following is an acute angle?
A. 65°
B. 90°
C. 125°
D. 180°

A. 65° is the answer. An acute angle is less than 90°.

8. Which of the following is an obtuse angle?
A. 65°
B. 90°
C. 125°
D. 180°

C. 125° is the answer. An obtuse angle is bigger than 90° and less than 180°.

9. Which of the following is a right angle?
A. 65°
B. 90°
C. 125°
D. 180°

B. 90° is the answer. A right angle is 90°.

Area, Circumference and Perimeter Review

Area - the space inside a shape
An example of area is how much carpet a room needs.

Square - all 4 sides are the same length and the shape has 4 right angles
Area = s^2 = side x side
If one side equals 4, the area is 4^2, which equals 16.
Perimeter of a square   P = 4s = 4 x length of side

Example:
What is the area of a square with a side of 8?
Area of a square   A = $side^2$ = side x side
Side = 8
Area = 8 x 8 =
Area = 64

Rectangle
Area = length x width
Perimeter of a rectangle = 2L + 2W = (2 x length) + (2 x width)

Example:
What is the area of a rectangle with a length of 4 and a width of 6?
Area of a rectangle   A = length x width
Length = 4, width = 6
Area = 4 x 6 =
Area = 24

Circle

Circumference
Circumference of a circle is the measure around the outside of the circle.
C = pi x diameter = π d
pi = π = 3.14
C = πd
<u>C</u>herry <u>p</u>ie is <u>d</u>elicious.

Area of a Circle
Area of a circle measures the inside of the circle.
A = πr^2
<u>A</u>pple <u>p</u>ies <u>a</u>re <u>t</u>oo.

Example:
What is the circumference of a circle with a radius of 6?

Circumference of a circle   C = pi x diameter = πd

Radius = 6, therefore diameter = 12

C = 3.14 x 12 =

Circumference = 37.68

The ASVAB is a multiple-choice test. Before solving a question with π (pi) in it, check the answer choices. If the answer choices are not too close in value, use 3 instead of 3.14 for π (pi). Solve quicker and choose the answer that is a little bigger than the answer you picked.
C = 3 x 12 = 36   The estimate is 36, but the actual answer is 37.68.

Triangle

Area = ½ x base x height

Perimeter = s1 + s2 + s3

All the angles of a triangle add up to 180 degrees.

What is the area of a triangle with a base of 8 and a height of 3?

Area of a triangle   A = ½ x base x height

base = 8, height = 3

Area = 1/2 x 8 x 3 = 4 x 3 =

Area = 12

Circle

What is the area of a circle with a radius of 7?

Area of a circle $A = \text{pi} \times \text{radius}^2 = \pi r^2$

Radius = 7

$A = 3.14 \times 7^2 = 3.14 \times 49 =$

Area = 153.86

When you plug in 3 instead of 3.14 for π (pi), the estimate is 147.

The answer must be a bit bigger than 147 with the actual answer being 153.86.

What is the area of a circle with a diameter of 10?

Area of a circle $A = \text{pi} \times \text{radius}^2 = \pi r^2$

To find area in this case, you need to change diameter to radius.

Radius is ½ the diameter. Since the diameter is 10, the radius is 5.

Radius = 5

$A = 3.14 \times 5^2 = 3.14 \times 25 =$

Area = 78.5

When you plug in 3 instead of 3.14 for π (pi), you get an estimate of 3 x 25 = 75, when the actual answer is 78.5

Perimeter - the measure of the outside of a shape
An example of perimeter is how much fence is needed.

The perimeter of any shape is found by adding up the lengths of all the sides.

Perimeter

What is the perimeter of a hexagon when each side equals 7 cm?

A hexagon has 6 sides.
Each side is 7 cm.
You could add 7 + 7 + 7 + 7 + 7 + 7 = 42 or do 6 x 7 = 42

Perimeter = 42 cm

Perimeter and Area Question in One Problem

If the perimeter of a rectangle is 36 and the length is 12 what is the area?

The perimeter of a rectangle is 36.

P = 2L + 2W
Length is 12
2L + 2W = 36
2(12) + 2W = 36
24 + 2W = 36
Subtract 24 from each side.
2W = 12
Divide by 2 on each side.

W = 6
Length is 12, and width is 6.

Area of a rectangle is length x width

Area = 12 x 6 = 72

Area of a Trapezoid

Area = 1/2(b1+b2)h

What is the area of a trapezoid with one base equal to 5, the other base equal to 7, and the height equal to 2?

Area of a trapezoid = 1/2 (base$_1$ + base$_2$)height
Base$_1$ = 5, Base$_2$ = 7, Height = 2
A = 1/2(5 + 7)2 = 1/2(12)2 = 6 x 2 =
Area = 12

Area of a Parallelogram

$A = bh$

What is the area of a parallelogram with a base of 10 and a height of 2?

Area of a parallelogram $A = bh$

base = 10, height = 2

$A = 10 \times 2 =$

Area = 20

Mathematics Knowledge Assessment
Area, Circumference, Perimeter - Pre-Test

1. What is the area of a square with a side of 6?

2. What is the area of a rectangle with a length of 7 and a width of 8?

3. What is the area of a triangle with a base of 6 and a height of 5?

4. What is the area of a parallelogram with a base of 8 and a height of 3?

5. What is the circumference of a circle with a radius of 3?

6. What is the area of a circle with a radius of 4?

7. What is the area of a circle with a diameter of 6?

8. What is the area of a trapezoid with one base equal to 6, the other base equal to 8 and the height equal to 4?

9. How do you find the perimeter of a shape?

10. What is the perimeter of a dodecagon when each side equals 9 cm?

11. If the perimeter of a rectangle is 54 and the length is 18, what is the area?

Julie A. Hyers © 2020

Mathematics Knowledge

Area, Circumference, Perimeter - Pre-Test with Answers

1. What is the area of a square with a side of 6?

    Area of a square   A = side^2 = side x side

    Side = 6

    Area = 6 x 6 =

    Area = 36

2. What is the area of a rectangle with a length of 7 and a width of 8?

    Area of a rectangle   A = length x width

    Length = 7, width = 8

    Area = 7 x 8 =

    Area = 56

3. What is the area of a triangle with a base of 6 and a height of 5?

    Area of a triangle   A = 1/2 x base x height

    base = 6, height = 5

    Area = 1/2 x 6 x 5 = 3 x 5 =

    Area = 15

4. What is the area of a parallelogram with a base of 8 and a height of 3?

    Area of a parallelogram   A = bh

    base = 8, height = 3, Area = 8 x 3 = 24

5. What is the circumference of a circle with a radius of 3?

Circumference of a circle   C = pi x diameter = πd

Radius = 3, therefore diameter = 6

C = 3.14 x 6 =

Circumference = 18.84

The ASVAB is a multiple-choice test. Before solving a question with π (pi) in it, check the answer choices. If the answer choices are not too close in value, use 3 instead of 3.14 for π (pi). Solve quicker and choose the answer that is a little bigger than the answer you picked.

C = 3 x 6 = 18   The estimate is 18, but the actual answer is 18.84.

6. What is the area of a circle with a radius of 4?

Area of a circle   A = pi x radius^2 = πr^2

Radius = 4

A = 3.14 x 4^2 = 3.14 x 16 =

Area = 50.24

When you plug in 3 instead of 3.14 for π (pi), the estimate is 48.

The answer must be a bit bigger than 48 with the actual answer being 50.24.

7. What is the area of a circle with a diameter of 6?

Area of a circle   A = pi x radius^2 = πr^2

To find area in this case, you need to change diameter to radius.

Radius is 1/2 the diameter. Since the diameter is 6, the radius is 3.

Radius = 3

$A = 3.14 \times 3^2 = 3.14 \times 9 =$

Area = 28.26

When you plug in 3 instead of 3.14 for π (pi), you get an estimate of 3 x 9 = 27, when the actual answer is 28.26.

8. What is the area of a trapezoid with one base equal to 6, the other base equal to 8 and the height equal to 4?

   Area of a trapezoid = 1/2(base₁ + base₂)height

   Base₁ = 6, Base₂ = 8, Height = 4
   A = 1/2(6 + 8)4 = 1/2(14)4 = 7 x 4 =
   Area = 28

9. How do you find the perimeter of a shape?

   Add up all the sides of a shape to find the perimeter.

10. What is the perimeter of a dodecagon when each side equals 9 cm?

    A dodecagon has 12 sides.

    Each side is 9 cm and 12 x 9 =
    Perimeter = 108 cm

11. If the perimeter of a rectangle is 54 and the length is 18, what is the area?

    The perimeter of a rectangle is 54.

    P = 2L + 2W

    Length is 18

    2L + 2W = 54

    2(18) + 2W = 54

    36 + 2W = 54

Subtract 36 from each side.

2W = 18

Divide by 2 on each side.

W = 9

Length is 18 and width is 9.

Area of a rectangle is length x width

Area = 18 x 9 = 162

Volume and Surface Area Review

Volume- the measure of how much space is inside a three-dimensional object
An example of volume is how much water will fill up a pool.

Volume of a Cylinder (can shape)

Volume = $\pi r^2 h$ = pi x radius^2 x height

Example for Cylinder:

What is the volume of a cylinder with a radius of 4 and a height of 2?

Volume of a cylinder = pi x radius^2 x height

V = 3.14 x 4^2 x 2 = 3.14 x 16 x 2 = 3.14 x 32 =

Volume = 100.48

If the answer choices on the test are spaced out, you can try to use 3 instead of 3.14 for pi to get

an estimate. The real answer will be a bit bigger than the estimate.

When you use 3 instead of 3.14 for pi, the answer is 3 x 4^2 x 2 = 3 x 16 x 2 = 96. The estimate is

96 when the actual answer is 100.48

Volume of a Sphere (Ball)

Volume = $4/3 \pi r^3$

This is a time-consuming formula for a timed test.
If the answer choices on the test are spaced out, you can try the shortcut formula.
Use 3 for π, when you do this, the new formula for sphere becomes simpler.
V = 4/3 x 3r^3 = 4/3 x 3/1r^3 = 4r^3

Simpler Sphere formula
V = 4r^3
The real answer will be a bit bigger than the answer found with the simpler formula, but it is a good estimate when the answer choices are not too close.

Example for Sphere:

What is the volume of a sphere with a radius of 2?

The real formula for volume of a sphere is not very simple and takes up time on a timed test.

Volume of a sphere = 4/3 x pi x radius^3

V = 4/3 x 3.14 x 2^3

V = 4/3 x 3.14 x 8 =

4/3 can be written as a decimal = 1.33

V = 1.33 x 3.14 x 8 =

This is also an estimate since 1.33 is a nonterminating decimal. The real answer is a little bigger too.

Volume = 33.4096

When you plug in 3 instead of 3.14 for pi, the answer is as follows,

V = 4/3 x 3 x radius^3 = 4/3 x 3/1 x radius^3

The two 3s cancel out, leaving the simpler formula, V = 4 x radius^3

The simpler sphere formula is V = 4r^3

V = 4 x 2^3 = 4 x 8 = 32. The estimate is 32 when the actual answer is 33.4096

The estimate can save time on the exam as long as the answer choices on the test are not too close in value. Make sure to pick the answer choice that is a little bigger than your estimate.

Volume of a Rectangular Box

V = L x W x H

Example for Volume of a Rectangular Box

What is the volume of a rectangular box with a length of 6, width of 4, and a height of 2?

Volume of a rectangular box = L x W x H

V = 6 x 4 x 2 =

Volume = 48

Volume of a Cube

V = side^3 = s^3

Example of Volume of a Cube

What is the volume of a cube with a side of 3?

Volume of a cube = side^3

V = s^3 = 3^3 = 3 x 3 x 3 =

Volume = 27

Surface Area- the measure of the total area of the surface of a three-dimensional object
An example of surface area would be how much wrapping paper is needed to wrap a present.

Surface Area of a Rectangular Box

SA = 2(LW + LH + WH)
SA = 2 [(Length x Width) + (Length x Height) + (Width x Height)]

Example of Surface Area of a Rectangular Box

What is the surface area of a rectangular box with a length of 3, width of 8, and a height of 10?

Surface area of a rectangular box = 2(LW + LH + WH)

L = 3, W = 8, H = 10

SA = 2[(3 x 8) + (3 x 10) + (8 x 10)] = 2 (24 + 30 + 80) = 2(134) =

Surface area = 268

Surface Area of a Cube

SA = $6s^2$ = 6 x side x side

Example of Surface Area of a Cube

What is the surface area of a cube with a side of 4?

Surface area of a cube = 6 x $side^2$

SA = $6(4)^2$ = 6 x 16 =

Surface area = 96

Mathematics Knowledge
Volume and Surface Area - Pre-Test

1. What is the volume of a cylinder with a radius of 5 and a height of 3?

2. What is the volume of a sphere with a radius of 3?

3. What is the volume of a rectangular box with a length of 4, width of 5, and a height of 6?

4. What is the volume of a cube with a side of 6?

5. What is the surface area of a rectangular box with a length of 2, width of 4, and a height of 5?

6. What is the surface area of a cube with a side of 7?

Julie A. Hyers © 2020

Mathematics Knowledge
Volume and Surface Area – Pre-Test with Answers

1. What is the volume of a cylinder with a radius of 5 and a height of 3?

   Volume of a cylinder = pi x radius^2 x height = $\pi r^2 h$

   V = 3.14 x 5^2 x 3 = 3.14 x 25 x 3 = 3.14 x 75 =
   Volume = 235.5

   If the answer choices on the test are spaced out, you can try to use 3 instead of 3.14 for pi to get an estimate. The real answer will be a bit bigger than the estimate.
   When you use 3 instead of 3.14 for pi, the answer is 3 x 5^2 x 3 = 3 x 25 x 3 = 225.
   The estimate is 225 when the actual answer is 235.5.

2. What is the volume of a sphere with a radius of 3?

   The real formula for volume of a sphere is not simple and takes up time on a timed test.

   Volume of a sphere = 4/3 x pi x radius^3 = $4/3 \pi r^3$

   V = 4/3 x 3.14 x 3^3
   V = 4/3 x 3.14 x 27 =
   4/3 can be written as a decimal = 1.33
   V = 1.33 x 3.14 x 27 =
   Volume = 112.7574

   When you plug in 3 instead of 3.14 for pi, the answer is as follows,
   V = 4/3 x 3 x radius^3 = 4/3 x 3/1 x radius^3
   The two 3s cancel out, leaving the simpler formula, V = 4 x radius^3
   The simpler sphere formula is V = 4r^3
   V = 4 x 3^3 = 4 x 27 = 108.

   The estimate is 108 when the actual answer is 112.7574

   The estimate can save time on the exam as long as the answer choices on the test are not too close in value. Pick the answer choice that is a little bigger than your estimate.

3. What is the volume of a rectangular box with a length of 4, width of 5, and a height of 6?

   Volume of a rectangular box = L x W x H

   V = 4 x 5 x 6 =
   Volume = 120

4. What is the volume of a cube with a side of 6?

   Volume of a cube = side^3

   V = s^3 = 6^3 = 6 x 6 x 6 =
   Volume = 216

5. What is the surface area of a rectangular box with a length of 2, a width of 4, and a height of 5?

   Surface area of a rectangular box = 2(LW + LH + WH)

   L = 2, W = 4, H = 5
   Surface Area = 2[(2 x 4) + (2 x 5) + (4 x 5)] = 2 (8 + 10 + 20) = 2(38) = 76

6. What is the surface area of a cube with a side of 7?

   Surface area of a cube = 6 x side^2

   Surface area = 6(7)^2 = 6 x 49 = 294

Real Life Applications of Perimeter, Area, Surface Area, Volume Review

Perimeter- the measure of the outside of a shape
An example of perimeter is how much fence is needed.
The perimeter of any shape is found by adding up the lengths of all the sides.
Perimeter is written as yards, feet, inches, miles, etc.

Example for perimeter:

How much fence is needed to enclose a backyard with dimension of 75 yd by 50 yd?

A question about how much fence is needed is a perimeter question. Perimeter involves adding up all the sides of a shape. A backyard with a dimension of 75 yd by 50 yd has 4 sides: 75 yd, 75 yd, 50 yd, and 50 yd.

For perimeter, add up the sides.
75 yd + 75 yd + 50 yd + 50 yd =
Perimeter = 250 yd

Perimeter is written as yards, feet, inches, miles, etc.

Area- the shape inside a shape
An example of area is how much carpet a room needs.
Area is written as square feet (ft^2), square inches ($in.^2$), square yards (yd^2), square miles (mi^2), etc., depending on the example.

Example for Area:

How much carpet is needed to cover the floor of a living room with dimensions of 30 ft x 12 ft?

A question about how much carpet is needed is an area question.

Area of a rectangle = length x width

30 x 12 = 360 square feet =

Area = 360 ft^2

Area is written as square feet (ft^2), square inches ($in.^2$), square yards (yd^2), square miles (mi^2), etc., depending on the example.

Surface area- the measure of the total area of the surface of a three-dimensional object
An example of surface area would be how much wrapping paper is needed to wrap a present.
Surface area is written as square feet (ft^2), square inches ($in.^2$), square yards (yd^2), square miles (mi^2), etc., depending on the example.

Example for Surface Area of a Square Box:

How much wrapping paper is needed to cover a square box with a height of 3 inches?

A question about wrapping paper is a question on surface area.

The surface area of a square box = 6 x side^2

$6(3)^2 = 6(9) = 54$ square feet =
Surface area = 54 ft^2

Surface area is written as square feet (ft^2), square inches (in.^{2}), square yards (yd^2), square miles (mi^2), etc., depending on the example.

Example for Surface Area of a Rectangular Box:

How much wrapping paper is needed to cover a rectangular box with a length of 3 inches, a width of 5 inches and a height of 6 inches?

A question about wrapping paper is a question on surface area.

The surface area of a rectangular box = 2[(length x width) + (length x height) + (width x height)]

Surface area of a rectangular box = 2(LW + LH + WH)

Length = 3
Width = 5
Height = 6

$2[3(5) + 3(6) + 5(6)] = 2(15 + 18 + 30) = 2(63) = 126$ square feet
Surface area =126 ft^2

Surface area is written as square feet (ft^2), square inches (in.^{2}), square yards (yd^2), square miles (mi^2), etc., depending on the example.

Volume- the measure of how much space is inside a three-dimensional object
An example of volume is how much water will fill up a pool.
Volume is written as cubic feet (ft^3), cubic inches (in.^{3}), cubic yards (yd^3), cubic miles (mi^3), etc., depending on the example.

Example of Volume of a Cylinder:

A round pool has a diameter of 10 ft and a height of 6 ft. A cubic foot of water is about 7.5 gallons. How many gallons are used to fill the pool?
A question about how much water is in a pool is a volume question.
The volume of a round pool is a question asking about the volume of a cylinder
Volume of a cylinder = π x radius² x height

$V = \pi r^2 h$

The problem gives the diameter of the pool.
Change diameter to radius to solve the problem.

If diameter is 10 ft, the radius is 5 ft.
The height is 6.

$V = \pi \times 5^2 \times 6 = \pi \times 25 \times 6 = 150\pi$
$V = 3.14 \times 150 = 471$ square feet $= 471$ ft³
The pool has a volume of 471 ft³

Volume is written as cubic feet (ft³), cubic inches (in.³), cubic yards (yd³), etc., depending on the example.

To figure out how many gallons of water are in the pool, multiply the volume of 471 ft³ by 7.5 since there are 7.5 gallons of water in each cubic foot of water.
471 x 7.5 = 3,532.5 gallons

Example of Volume of a Rectangular Box:

A rectangular pool has dimensions of 7 ft by 15 ft x 60 ft. A cubic foot of water is about 7.5 gallons. How many gallons are used to fill the pool?

A question about how much water is in a pool is a volume question.
The volume of a rectangular pool is length x width x height.

Volume = L x W x H = 7 x 15 x 60 = 6,300 cubic feet = 6,300 ft³

Volume is written as cubic feet (ft³), cubic inches (in.³), cubic yards (yd³), etc. depending on the example.

To figure out how many gallons of water are in the pool, multiply the volume of 6,300 ft² by 7.5 since there are 7.5 gallons of water in each cubic foot of water.
6,300 x 7.5 = 47,250 gallons

Mathematics Knowledge
Real Life Applications for Perimeter, Area, Surface Area, Volume - Pre-Test

1. How much fence is needed to enclose a backyard with dimensions of 100 yd by 200 yd?

2. How much carpet is needed to cover the floor of a living room with dimensions of 40 ft x 15 ft?

3. How much wrapping paper is needed to cover a square box with a height of 5 inches?

4. How much wrapping paper is needed to cover a rectangular box with a length of 6 inches, a width of 8 inches and a height of 3 inches?

5. A round pool has a diameter of 30 ft and a height of 5 ft. A cubic foot of water is about 7.5 gallons. How many gallons are used to fill the pool?

6. A rectangular pool has dimensions of 15 ft by 20 ft by 80 ft. A cubic foot of water is about 7.5 gallons. How many gallons are used to fill the pool?

Julie A. Hyers © 2020

Mathematics Knowledge
Real Life Applications for Perimeter, Area, Surface Area, Volume - Pre-Test with Answers

1. How much fence is needed to enclose a backyard with dimensions of 100 yd by 200 yd?

   A question about how much fence is needed is a perimeter question. Perimeter involves adding up all the sides of a shape. A backyard with a dimension of 100 yd by 200 yd has 4 sides: 100 yd, 100 yd, 200 yd, and 200 yd.

   For perimeter, add up the sides.
   100 yd + 100 yd + 200 yd + 200 yd =
   Perimeter = 600 yd

   Perimeter is written as yards, feet, inches, miles, etc.

2. How much carpet is needed to cover the floor of a living room with dimensions of 40 ft x 15 ft?

   A question about how much carpet is needed is an area question.

   Area of a rectangle = length x width

   40 x 15 = 600 square feet =
   Area = 600 ft^2

   Area is written as square feet (ft^2), square inches (in.^{2}), square yards (yd^2), square miles (mi^2), etc., depending on the example.

3. How much wrapping paper is needed to cover a square box with a height of 5 inches?

   A question about wrapping paper is a question on surface area.

   The surface area of a square box = 6 x side^2

   $6(5)^2 = 6(25) = 150$ square feet =

   Surface area = 150 ft^2

   Surface area is written as square feet (ft^2), square inches (in.^{2}), square yards (yd^2), square miles (mi^2), etc., depending on the example.

4. How much wrapping paper is needed to cover a rectangular box with a length of 6 inches, a width of 8 inches, and a height of 3 inches?

A question about wrapping paper is a question on surface area.

The surface area of a rectangular box =
2[(length x width) + (length x height) + (width x height)]

Surface area of a rectangular box = 2(LW + LH + WH)

Length = 6
Width = 8
Height = 3

2[6(8) + 6(3) + 8(3)] = 2 (48 + 18 + 24) = 2(90) = 180 square feet =
Surface area =180 ft^2

Surface area is written as square feet (ft^2), square inches (in.^{2}), square yards (yd^2), square miles (miles^2), etc., depending on the example.

5. A round pool has a diameter of 30 ft and height of 5 ft. A cubic foot of water is about 7.5 gallons. How many gallons are used to fill the pool?

A question about how much water is in a pool is a volume question.
The volume of a round pool is a question asking about the volume of a cylinder.
Volume of a cylinder = π x radius^2 x height

V = πr^{2}h

The problem gives the diameter of the pool.
Change diameter to radius to solve the problem.

If diameter is 30 ft, the radius is 15 ft.

V = π x 15^2 x 5 = π x 225 x 5 = 1125
V = 3.14 x 1125 = 3532.5 square feet = 3532.5 ft^3
The pool has a volume of 3,532 ft^3.

Volume is written as cubic feet (ft^3), cubic inches (in.^{3}), cubic yards (yd^3), etc., depending on the example.

To figure out how many gallons of water are in the pool, multiply the volume of 3532.5 ft^2 by 7.5 since there are 7.5 gallons of water in each cubic foot of water.
3532.5 x 7.5 = 26,493.75 gallons

6. A rectangular pool has dimensions of 15 ft by 20 ft by 80 ft. A cubic foot of water is about 7.5 gallons. How many gallons are used to fill the pool?

A question about how much water is in a pool is a volume question.
The volume of a rectangular pool is length x width x height.

Volume = 15 x 20 x 80 = 24,000 cubic feet = 24,000 ft^3

Volume is written as cubic feet (ft^3), cubic inches (in.^{3}), cubic yards (yd^3), etc. depending on the example.

To figure out how many gallons of water are in the pool, multiply the volume of 24,000 ft.^{2} by 7.5 since there are 7.5 gallons of water in each cubic foot of water.
24,000 x 7.5 = 180,000 gallons

PEMDAS, Scientific Notation, Factorial Review

Order of Operations- PEMDAS

Parentheses
Exponents
Multiplication or Division - whichever comes first
Addition or Subtraction - whichever comes first

Example:

$8 + 2 - (7-2)^2 \times 2 + 12 \div 4 =$

$8 + 2 - (5)^2 \times 2 + 12 \div 4 =$

$8 + 2 - 25 \times 2 + 12 \div 4 =$

$8 + 2 - 50 + 12 \div 4 =$

$8 + 2 - 50 + 3 =$

$10 - 50 + 3 =$

$-40 + 3 = -37$

Scientific Notation is a way to write very large or very small numbers.

To be written in scientific notation, the decimal place must be placed after the first digit in a number.
The number is multiplied by 10 to a power that shows how many spaces the decimal point moved over.

If the number that needs to be changed into scientific notation is a whole number, the exponent will be positive.

If the number that needs to be changed into scientific notation is a decimal (too small to be a whole number), the exponent will be negative.

Examples:

1,456,873 written in scientific notation is 1.456873×10^6.
The 6 means that in order to return to the original number, the decimal is moved 6 places to the right.

0.05921 written in scientific notation is 5.921×10^{-2}.
The -2 means that in order to return to the original number, the decimal is moved 2 places to the left.

Factorial is the product of a number and all the numbers below it down to 1.
It is used in problems with combinations.
Factorial can be written in 2 ways: a number with the word factorial after it or a number with an exclamation point after it.

Example: Factorial can be written as 5 factorial or 5!
Both ways mean 5 x 4 x 3 x 2 x 1 = 120

Another example:
If there are 4 books on a shelf, how many ways can the books be arranged?

4 factorial can also be written as 4!
It means 4 x 3 x 2 x 1 = 24
There are 24 ways the 4 books can be arranged on the shelf.

A tricky example is 0 factorial or 0!
It equals 1. Remember this.
The explanation of why 0 factorial or 0! is equal to one is this,
If you have 0 books, how many ways can you arrange the 0 books on the shelf.
There is one way, an empty shelf!

Mathematics Knowledge
PEMDAS, Scientific Notation, Factorial - Pre-Test

1. $5 + 1 - (8 - 4)^2 \times 3 + 9 \div 3 =$

2. Write in scientific notation.
   2,831,834

3. Write in scientific notation.
   0.00526

4. Write as a numeral.
   9.165×10^2

5. Write as a numeral.
   3.7862×10^{-3}

6. Solve.
   4!

7. Solve.
   0!

8. Solve.
   6 factorial

Mathematics Knowledge
PEMDAS, Scientific Notation, Factorial - Pre-Test with Answers

1. $5 + 1 - (8-4)^2 \times 3 + 9 \div 3 =$

    Follow PEMDAS
    Work on Parentheses first.
    Then work on Exponents.
    Solve Multiplication and Division whichever comes first in left to right order.
    Solve Addition and Subtraction whichever comes first in left to right order.

    $5 + 1 - (8-4)^2 \times 3 + 9 \div 3 =$

    $5 + 1 - (4)^2 \times 3 + 9 \div 3 =$

    $5 + 1 - 16 \times 3 + 9 \div 3 =$

    $5 + 1 - 48 + 3 =$

    $6 - 48 + 3 =$

    $-42 + 3 =$

    -39

2. Write in scientific notation.
    2,831,834

    2.831834×10^6

3. Write in scientific notation.
    0.00526

    5.26×10^{-3}

4. Write as a numeral.
    9.165×10^2

    916.5

5. Write as a numeral.
   3.7862 x 10^{-3}

   0.0037862

6. Solve.
   4!

   4 x 3 x 2 x 1 = 24

7. Solve.
   0!

   0! = 1

8. Solve.
   6 factorial

   6 x 5 x 4 x 3 x 2 x 1 = 720

Exponents Review

When working with exponents, be sure to follow the rules of **PEMDAS.**

Order of Operations

Parentheses
Exponents
Multiplication or Division whichever comes first
Addition or Subtraction whichever comes first

Evaluate

$3x^4 + 2y^3$
when $x = 1$, $y = -3$

$3(1)^4 + 2(-3)^3 =$
Follow order of operations (PEMDAS)
$3(1) + 2(-27)$
$3 - 54 = -51$

Evaluate

$4a^2 + 5b^3 - c^3$
when $a = -2$, $b = 3$, $c = -1$

Follow order of operations (PEMDAS)
$4(-2)^2 + 5(3)^3 - (-1)^3 =$
$4(4) + 5(27) - (-1) =$
$16 + 135 + 1 = 152$

Division with Exponents

When dividing with exponents, subtract the exponents.
$$\frac{10x^4y^6z^7}{5x^2y^8z^4}$$

Divide or reduce the numerals and subtract the exponents.
$2x^2y^{-2}z^3$
If there is a variable with a negative exponent, it must be moved down to the denominator to become positive.
The final answer is $\dfrac{2x^2z^3}{y^2}$

Raising Exponents to a Power

$(4x^4y^5)^3 =$

Raise the number 4 to the 3rd power. $4^3 = 64$
and multiply the exponents by 3
$64x^{12}y^{15}$

Raising Exponents to a Power and Dividing Exponents

$$\frac{(3x^3y^5)^3}{9x^5y^8} =$$

Work on the exponent first.
Raise 3 to the 3rd power and multiply the exponents by 3 in the numerator.

$$\frac{27x^9y^{15}}{9x^5y^8} =$$

Then divide or reduce the numbers and subtract the exponents.

$3x^4y^7$

Multiplying Exponents

$(3x^6)(5x^8) =$

Multiply the numbers and add the exponents.

$15x^{14}$

Evaluating with Exponents

Evaluate when $x = -2$

$$\frac{4x^6}{x^3} =$$

Simplify by dividing x^3, which leads to $4x^3$.

$$\frac{4x^6}{x^3} = 4x^3 = 4(-2)^3 = 4(-8) = -32$$

Julie A. Hyers © 2020

Mathematics Knowledge
Exponents - Pre-Test

1. Evaluate.
   $5x^3 + 2y^4$
   when $x = 2$, $y = -2$

   $5(2)^3 + 2 \cdot (-2)^4$
   $5 \cdot 8 + 2 \cdot 16 =$
   $40 + 32 = 72$

2. Evaluate.
   $5a^2 + 6b^3 - 2c^3$
   when $a = -1$, $b = 2$, $c = -3$

   $5(-1)^2 + 6(2)^3 - 2(-3)^3$
   $5 \cdot 1 + 6 \cdot 8 - 2(-27)$
   $5 + 48 + 52 = 105$

3. $\dfrac{6x^3y^4z^5}{3x^2y^5z^4} =$

   $\dfrac{2 \cdot x^{(3-2)} \cdot z^{(5-4)}}{y^{(5-4)}} = \dfrac{2 \cdot x \cdot z}{y}$

   36
   × 6
   216

4. $(6x^3y^4)^3 =$

   $216 \cdot x^9 \cdot y^{12}$

5. $\dfrac{(2x^2y^4)^3}{8x^4y^9} = \dfrac{(2x^2y^4)^3}{8x^4y^9} = \dfrac{8 \cdot x^6 \cdot y^{12}}{8x^4y^9} = x^2 \cdot y^3$

   $2 \times 2 \times 2$

6. $(2x^4)(3x^6) =$

$6x^{(4+6)}$

7. Evaluate when $x = 2$

$\dfrac{3x^5}{x^2} = \dfrac{3(2)^5}{2^2} \quad \dfrac{3 \times \cancel{32}^{\;8}}{\cancel{4}} = 24$

Julie A. Hyers © 2020

Mathematics Knowledge
Exponents - Pre-Test with Answers

1. Evaluate.
   $5x^3 + 2y^4$
   when $x = 2$, $y = -2$

   $5(2)^3 + 2(-2)^4 =$
   Follow order of operations (PEMDAS)
   $5(8) + 2(16)$
   $40 + 32 = 72$

2. Evaluate.
   $5a^2 + 6b^3 - 2c^3$
   when $a = -1$, $b = 2$, $c = -3$

   Follow order of operations (PEMDAS)
   $5(-1)^2 + 6(2)^3 - 2(-3)^3$
   $5(1) + 6(8) - 2(-27) =$
   $5 + 48 + 54 = 107$

3. $\dfrac{6x^3y^4z^5}{3x^2y^5z^4} =$

   Divide or reduce the numerals and subtract the exponents.
   $2xy^{-1}z$
   If there is a variable with a negative exponent, it must be moved down to the denominator to become positive.
   The final answer is $\dfrac{2xz}{y}$

4. $(6x^3y^4)^3 =$

   Raise the number 6 to the 3rd power. $6^3 = 216$
   Multiply the exponents by 3.
   $216x^9y^{12}$

5. $\dfrac{(2x^2y^4)^3}{8x^4y^9} =$

Work on the exponent first.
Raise 2 to the 3rd power and multiply the exponents by 3 in the numerator.

$\dfrac{8x^6y^{12}}{8x^4y^9}$

Then divide or reduce the numbers and subtract the exponents.
x^2y^3

6. $(2x^4)(3x^6) =$
Multiply the numbers and add the exponents.
$6x^{10}$

7. Evaluate when $x = 2$
$\dfrac{3x^5}{x^2} = 3x^3 = 3(2)^3 = 3(8) = 24$

Julie A. Hyers © 2020

Factoring and FOIL (Distribution) in Algebra Review

Algebra

FOIL stands for First, Outer, Inner, Last
It stands for the steps involved in distribution in algebra.
$(x + 2)(x + 3)$
$x^2 + 3x + 2x + 6$
Combine like terms $3x + 2x = 5x$
$x^2 + 5x + 6$

Factoring Rules in Algebra

First Scenario

$x^2 + 10x + 24$

If the second sign is positive, it means both signs are the same when you factor.
If the first sign is positive, both signs are positive.
You need to find what numbers add up to the first number and multiply to the second number.

$x^2 + 10x + 24$

What 2 numbers add up to +10 and multiply to +24?

$\quad 6 + 4 = 10$ and $6 \times 4 = 24$

Therefore, the answer is:

$(x + 6)(x + 4)$

Second Scenario

$x^2 - 12x + 35$

If the second sign is positive, it means both signs are the same when you factor.
If the first sign is negative, it means both signs are negative.
You need to find what numbers add up to the first number and multiply to the second number.

$x^2 - 12x + 35$

What 2 numbers add up to -12 and multiply to +35?

-7 + -5 = -12

and -7 x -5 = 35

Therefore, the answer is:

(x - 7)(x - 5)

Third Scenario

$x^2 + 2x - 15$

If the second sign is negative, both signs are different when you factor.
If the first sign is positive, the larger number is positive.
You need to figure out what numbers subtract to give you the first number and multiply to give you the second number.

$x^2 + 2x - 15$

What 2 numbers subtract to +2 and multiply to -15?
5 - 3 = 2 and 5 x -3 = -15

Therefore, the answer is
(x + 5)(x - 3)

Fourth Scenario

$x^2 - 4x - 21$

If the second sign is negative, both signs are different when you factor.
If the first sign is negative, the larger number is negative.
You need to figure out what numbers subtract to give you the first number and multiply to give you the second number.

$x^2 - 4x - 21$

What 2 numbers subtract to -4 and multiply to -21?

-7 + 3 = -4 and -7 x 3 = -21

Therefore, the answer is:

(x - 7)(x + 3)

Fifth Scenario

This case is called the difference of two squares.
This case is an example of a perfect square, subtraction, and another perfect square.

$x^2 - 25$

Find the square root of x^2 and the square root of 25.
The square root of x^2 is x, and the square root of 25 is 5.
The answers will be one positive and one negative.
$(x + 5)(x - 5)$

Another example is when the there is a digit before x^2

$9x^4 - 9x^2 - 10$

The second sign is negative which means the two signs will be different when you factor.
The first sign is negative which means the bigger number will be negative.
The last 2 digits must multiply to -10.

$(3x^2 - 5)(3x^2 + 2)$

FOIL (First Outer Inner Last)

Example:
$(x - 5)(x + 5)$

$x^2 + 5x - 5x - 25$

$x^2 - 25$

Example:
$(x + 3)(x + 10)$

$x^2 + 10x + 3x + 30$

$x^2 + 13x + 30$

Example:
$(x + 6)(x - 3)$

$x^2 - 3x + 6x - 18$

$x^2 + 3x - 18$

Example:
$(x - 11)(x - 4)$

$x^2 - 4x - 11x + 44$

$x^2 - 15x + 44$

Example:
$(3x + 2)(4x - 4)$

$12x^2 - 12x + 8x - 8$

$12x^2 - 4x - 8$

Mathematics Knowledge
Factoring and FOIL (Distribution) in Algebra - Pre-Test

1. $x^2 + 9x + 8$

    8   1

    $x+8$ $x+1$

2. $x^2 - 12x + 35$

    -7 -5

3. $x^2 - 4x - 32$

    -8 4

    $(x-8) \cdot (x+4)$

4. $x^2 + 9x - 36$

    -3 $+12$ = 9

    3 -12 = -9

5. $x^2 - 100$

    -10 $+10$

    $(x-10)(x+10)$

6. $4x^4 + 2x^2 - 6$

7. $(x-7)(x+7)$

    $x^2 + 7x - 7x - 49$

    $x^2 - 49$

8. $(x+2)(x+9)$

$x^2 + 11x + 18$

9. $(x+5)(x-1)$ $x^2 - x + 5x - 5$

$x^2 + 4x - 5$

10. $(x-6)(x-3)$ $-6 -3$

$x^2 - 3x - 6x + 18$

$x^2 - 9x + 18$

11. $(2x+5)(3x-2)$

$6x^2 - 4x + 15x - 10$

$-4x + 15x$

$11x$

$6x^2 + 11x - 10$

Mathematics Knowledge
Factoring and FOIL (Distribution) in Algebra - Pre-Test with Answers

1. $x^2 + 9x + 8$

    $(x + 8)(x + 1)$

2. $x^2 - 12x + 35$

    $(x - 7)(x - 5)$

3. $x^2 - 4x - 32$

    $(x - 8)(x + 4)$

4. $x^2 + 9x - 36$

    $(x + 12)(x - 3)$

5. $x^2 - 100$

    $(x + 10)(x - 10)$

6. $4x^4 + 2x^2 - 6$

    $(2x^2 + 3)(2x^2 - 2)$

7. $(x - 7)(x + 7)$

$x^2 + 7x - 7x - 49$

$x^2 - 49$

8. $(x + 2)(x + 9)$

$x^2 + 9x + 2x + 18$

$x^2 + 11x + 18$

9. $(x + 5)(x - 1)$

$x^2 - x + 5x - 5$

$x^2 + 4x - 5$

10. $(x - 6)(x - 3)$

$x^2 - 3x - 6x + 18$

$x^2 - 9x + 18$

11. $(2x + 5)(3x - 2)$

$6x^2 - 4x + 15x - 10$

$6x^2 + 11x - 10$

Solve for x with Distribution

5(x + 3) = 15(x + 1)

Work to get x on one side of the equal sign and the numbers on the other side of the equal sign.
Distribute 5(x + 3) = 5x +15
Distribute 15(x + 1) = 15x + 15

5x + 15 = 15x + 15
Subtract 5x from each side of the equal sign.
15 = 10x +15
Subtract 15 from each side.
10x = 0
Divide each side by 10.
x = 0

In this case, x = 0. This example can be confusing at times.
It might seem like there is no solution, but there is.

The solution is x = 0.

Solve for x with Distribution

7(x + 4) = 7(x + 3)

Distribute 7(x + 4) = 7x + 28
Distribute 7(x + 3) = 7x + 21

7x + 28 = 7x + 21
Subtract 7x from each side.
28 ≠ 21

This is an example where there is no solution. This might be confusing example, but remember it is possible to have no solution to a problem. No solution is the case here.

28 cannot equal 21.
There is no solution.

Another example that might seem tricky:

Just solve, step by step and you will find the value of x.
6 − x = x − 6, then x =
Add 6 to both sides.
12 − x = x
Add x to both sides.

12 = 2x
Divide by 2.

6 = x

Solving for x with Inequalities

8x – 16 ≤ 24

Add 16 to both sides.
8x ≤ 40
Divide by 8.

x ≤ 5

Solve for x with Inequalities

-2x + 7 ≥ 19

Subtract 7 from both sides.
-2x ≥ 12
Divide by -2.

When you divide by a negative number when dealing with an inequality, flip the sign around.

x ≤ -6

Solve for x with Inequalities

-9x – 2 ≤ -29

Add 2 to both sides.
 -9x ≤ -27
Divide by -9.

When you divide by a negative number when dealing with an inequality, flip the sign around.

x ≥ 3

Turning a Word Problem into an Algebra Problem Where You Solve for X

Ten less than 6 times a number is 32. What is the number?

Translate this problem into an algebraic equation.
Let the number equal x and solve for x.

6x – 10 = 32

Add 10 to both sides.
6x = 42
Divide by 6.

x = 7

Mathematics Knowledge
Solve for x - Pre-Test

Solve for x.

1. $5x + 3y = 19$
    $4x - 6y = 32$

    Handwritten work:
    $2x - 3y = 16$
    $2x - 3x = 16$
    $2x - 16 = 3y$
    $5x + (2x - 16) = 19$
    $7x - 16 = 19$
    $7x = 35 \rightarrow \underline{x = 5}$

2. $4x - 8 = 3x - 10$

3. $5(2x + 6) = 2(4x - 6)$

4. $3(4x + 2) = 2(2x + 3)$

5. $8(x + 1) = 8(x + 2)$

6. $5 - x = x - 5$, then $x =$

7. $3x - 6 \leq 12$

8. $-4x + 5 \geq 21$

9. $-5x - 2 \leq -32$

10.   Five less than 4 times a number is 35. What is the number?

Julie A. Hyers © 2020

Mathematics Knowledge
Solve for x - Pre-Test with Answers

Solve for x.

1. $5x + 3y = 19$
   $4x - 6y = 32$

   In order to solve for x, we need to work to eliminate y.

   To get rid of y, we want to make sure the numbers before y in the 2 equations are equal with opposite signs.

   We can change the first equation by multiplying the whole thing by 2.
   $2(5x + 3y = 19) = 10x + 6y = 38$

   Now if we combine the two equations through addition, the +6y and -6y will cancel out.

   $10x + 6y = 38$
   $+4x - 6y = 32$
   $14x = 70$
   $x = 5$

   If you need to also solve for y, plug x = 5 back into the easier equation.
   $5x + 3y = 19$
   $5(5) + 3y = 19$
   $25 + 3y = 19$
   $3y = -6$
   $y = -2$

2. $4x - 8 = 3x - 10$

   Work to get x on one side of the equal sign and the numbers on the other side of the equal sign.
   $4x - 8 = 3x - 10$
   Subtract 3x from both sides.
   $x - 8 = -10$
   Then add 8 to both sides.
   $x = -2$

3. $5(2x + 6) = 2(4x - 6)$

    Work to get the x on one side of the equal sign and the numbers on the other side of the equal sign.

    Distribute $5(2x + 6) = 10x + 30$
    Distribute $2(4x - 6) = 8x - 12$

    $10x + 30 = 8x - 12$
    Subtract 8x from each side.
    $2x + 30 = -12$
    Subtract 30 from each side.
    $2x + 30 = -12$
    $2x = -42$
    Divide by 2.
    $x = -21$

4. $3(4x + 2) = 2(2x + 3)$

    Work to get x on one side of the equal sign and the numbers on the other side of the equal sign.
    Distribute $3(4x + 2) = 12x + 6$
    Distribute $2(2x + 3) = 4x + 6$

    $12x + 6 = 4x + 6$
    Subtract 4x from each side of the equal sign.
    $8x + 6 = 6$
    Subtract 6 from each side.
    $8x = 0$
    Divide each side by 8.
    $x = 0$

5. $8(x + 1) = 8(x + 2)$

    Distribute $8(x + 1) = 8x + 8$
    Distribute $8(x + 2) = 8x + 16$
    $8x + 8 = 8x + 16$
    Subtract 8x from each side.
    $8 \neq 16$
    8 cannot equal 16.

    There is no solution.

6. $5 - x = x - 5$, then $x =$

     Add 5 to both sides.
     $10 - x = x$
     Add x to both sides.
     $10 = 2x$
     Divide by 2.
     $5 = x$

7. $3x - 6 \leq 12$

     Add 6 to both sides.
     $3x \leq 18$
     Divide by 3.
     $x \leq 6$

8. $-4x + 5 \geq 21$

     Subtract 5 from both sides.

     $-4x \geq 16$
     Divide by -4.

     When you divide by a negative number when dealing with an inequality, flip the sign around.

     $x \leq -4$

9. $-5x - 2 \leq -32$

     Add 2 to both sides.
     $-5x \leq -30$

     Divide by -5. When you divide by a negative number when dealing with an inequality, flip the sign around.

     $x \geq 6$

10. Five less than 4 times a number is 35. What is the number?

    Translate this problem into an algebraic equation.
    Let the number equal n and solve for n.
    (You can use any letter as a variable. It does not always need to be the letter x.)

    $4n - 5 = 35$

    Add 5 to both sides.
    $4n = 40$
    Divide by 4.

    $n = 10$

Julie A. Hyers © 2020

Basic Operations in Algebra Review

Addition with Polynomials

$$\begin{array}{r} 4x^2 - 5x + 3 \\ +\ \underline{x^2 + 2x - 6} \end{array}$$

To solve for addition with polynomials, just add up each term. Pay close attention to the signs.

$$\begin{array}{r} 4x^2 - 5x + 3 \\ +\ \underline{x^2 + 2x - 6} \\ 5x^2 - 3x - 3 \end{array}$$

Subtraction with Polynomials

$$\begin{array}{r} 5x^2 + 4x - 6 \\ -\ \underline{x^2 - 3x + 8} \end{array}$$

To solve for subtraction with polynomials, change subtraction to addition and change all the signs in the second expression.
Place parentheses around the second expression and change all the signs inside the parentheses.

$$\begin{array}{r} 5x^2 + 4x - 6 \\ +\ \underline{(-x^2 + 3x - 8)} \\ 4x^2 + 7x - 14 \end{array}$$

Multiplication with Polynomials

$$\frac{x^2}{x+7} \cdot \frac{x^2 + 11x + 28}{x} =$$

Factor the expression $x^2 + 11x + 28$ before multiplying the terms. Reduce diagonally if possible. Then multiply the numerators and the denominators straight across.

$$\frac{x^2}{x+7} \cdot \frac{(x+7)(x+4)}{x}$$

$$\frac{x^{2\,1}}{\cancel{x+7}} \cdot \frac{\cancel{(x+7)}(x+4)}{\cancel{x}}$$

$$x(x+4)$$

Division with Polynomials

$$\frac{x^2 - 36}{x^2} \div \frac{x + 6}{x^4} =$$

Factor the expression $x^2 - 36$. Since this is division of polynomials in fractions, be sure to keep, change, and flip. Keep the first fraction, change division to multiplication, flip the second fraction over. Reduce diagonally if possible. Multiply the numerators and the denominators straight across.

$$\frac{(x + 6)(x - 6)}{x^2} \cdot \frac{x^4}{x + 6}$$

$$\frac{\cancel{(x + 6)}(x - 6)}{x^2} \cdot \frac{x^{\cancel{4}\ 2}}{\cancel{x + 6}}$$

$$x^2(x - 6)$$

Mathematics Knowledge
Basic Operations in Algebra - Pre-Test

1.
$$3x^2 - 4x + 2$$
$$+\ \underline{3x^2 - 2x - 5}$$

2.
$$4x^2 + 5x - 6$$
$$-\ \underline{x^2 - 3x + 10}$$

3.
$$\frac{x^2}{x+3} \cdot \frac{x^2 + 6x + 9}{x} =$$

4.
$$\frac{x^2 - 16}{x^2} \div \frac{x - 4}{x^4} =$$

Mathematics Knowledge
Basic Operations in Algebra - Pre-Test with Answers

1.
$$\begin{array}{r} 3x^2 - 4x + 2 \\ +\ \underline{3x^2 - 2x - 5} \end{array}$$

To solve for addition with polynomials, just add up each term. Pay close attention to the signs.

$$\begin{array}{r} 3x^2 - 4x + 2 \\ +\ \underline{3x^2 - 2x - 5} \\ 6x^2 - 6x - 3 \end{array}$$

2.
$$\begin{array}{r} 4x^2 + 5x - 6 \\ -\ \underline{x^2 - 3x + 10} \end{array}$$

To solve for subtraction with polynomials, change subtraction to addition and change all the signs in the second expression. Place parentheses around the second expression and change all the signs inside the parentheses.

$$\begin{array}{r} 4x^2 + 5x - 6 \\ +\ \underline{(-x^2 + 3x - 10)} \\ 3x^2 + 8x - 16 \end{array}$$

3.
$$\frac{x^2}{x+3} \cdot \frac{x^2 + 6x + 9}{x} =$$

Factor the expression $x^2 + 6x + 9$ before multiplying the terms. Reduce diagonally if possible. Then multiply the numerators and the denominators straight across.

$$\frac{x^2}{x+3} \cdot \frac{(x+3)(x+3)}{x}$$

$$\frac{x^{2\ 1}}{\cancel{x+3}} \cdot \frac{(x+3)\cancel{(x+3)}}{\cancel{x}}$$

$$x(x+3)$$

105

4.

$$\frac{x^2 - 16}{x^2} \div \frac{x - 4}{x^4} =$$

Factor the expression x^2-16. Since this is division of polynomials in fractions, be sure to keep, change, flip. Keep the first fraction, change division to multiplication, flip the second fraction over. Reduce diagonally if possible. Multiply the numerators and the denominators straight across.

$$\frac{(x+4)(x-4)}{x^2} \cdot \frac{x^4}{x-4}$$

$$\frac{(x+4)\cancel{(x-4)}}{\cancel{x^2}} \cdot \frac{x^{4\,2}}{\cancel{x-4}}$$

$$x^2(x+4)$$

Algebra with Fractions Review

Algebra with Fractions

$3/8x = 24$

To get rid of a fraction in front of a variable, multiply by the reciprocal on both sides. Multiplying by the reciprocal cancels out the fraction in front of the variable.

$8/3 \bullet 3/8\ x = 24/1 \bullet 8/3$
$x = 192/3$
$x = 64$

Algebra with Fractions and Subtraction

$6/2x - 4 = 14$

Add 4 to both sides of the equation.
$6/2x - 4 + 4 = 14 + 4$
$6/2x = 18$
Multiply by the reciprocal on both sides.
Multiplying by the reciprocal cancels out the fraction in front of the variable.

$2/6 \bullet 6/2\ x = 18/1 \bullet 2/6$
$x = 36/6$
$x = 6$

Algebra with Fractions and Addition

$9/4x + 3 = 48$

Subtract 3 from both sides.

$9/4x + 3 - 3 = 48 - 3$
$9/4x = 45$

Multiply by the reciprocal on both sides.
Multiplying by the reciprocal on both sides of the equation cancels out the fraction in front of the variable.

$4/9 \bullet 9/4\ x = 45/1 \bullet 4/9$
$x = 45/1 \bullet 4/9$
Reduce the 45 and 9 to become 5 and 1.
$x = 5/1 \bullet 4/1 = 20/1$
$x = 20$

Addition with Fractions in Algebra

$1/6x + 1/7x =$

Add the 2 fractions.
Find a common denominator, and whatever you do to the denominator, you do to the numerator.
The least common denominator is 42.
$1/6 = 7/42$
and $1/7 = 6/42$
$7/42x + 6/42x = 13/42x$

Subtraction with Fractions in Algebra

$6/8x - 4/8x = 40$

This problem has a common denominator so you can subtract.
$2/8x = 40$

To get rid of the fraction in front of the variable, multiply by the reciprocal.

$8/2 \cdot 2/8 \, x = 40/1 \cdot 8/2 =$
Reduce 40 and 2 to become 20 and 1
$x = 20/1 \cdot 8/1 =$

$x = 160$

Solving for a Variable

Solve for z when $x = 4$
$$z = \frac{x^4}{4} - 16$$

Plug 4 in for x.
$$z = \frac{(4)^4}{4} - 16$$

$$z = \frac{256}{4} - 16 =$$

$z = 64 - 16 =$

$z = 48$

Using a Proportion to Solve for a Variable

If x/11 = 5/y, what is the value of xy?

Cross multiply to solve.

x/11 = 5/y

xy = 55

$$\frac{1}{2} \times \frac{2}{3} = \frac{1 \times 2}{2 \times 3} = \frac{2}{6}$$

$$\frac{4}{5} \cdot 5 = \frac{4 \cdot 5}{5}$$

$$ax + b = c$$
$$b(ax) = c$$

Mathematics Knowledge
Algebra with Fractions - Pre-Test

1. $4/5x = 20$

    $\frac{4}{5}x = 20 \rightarrow \frac{4}{5}x = 20 \Rightarrow \frac{4x}{5} = \frac{20}{1}$

    $4x = 100 \qquad\qquad\qquad 4x = 20 \cdot 5$
    $x = 25 \qquad\qquad\qquad\quad 4x = 100$
    $\qquad\qquad\qquad\qquad\qquad\quad x = 25$

2. $5/2x - 3 = 7$

    $\frac{5}{2}x - 3 = 7 \qquad\qquad \frac{5x}{2} = \frac{10}{1}$

    $\frac{5x}{2} - 3 = 7 \quad {+3 \atop +3} \qquad 20 = 5x$
    $\qquad\qquad\qquad\qquad\qquad\;\; \overline{\;5\;} \quad 5$
    $\frac{5x}{2} = 7 + 3 \qquad\qquad\quad x = 4$

3. $7/3x + 4 = 32$

    $\frac{7}{3}x + 4 = 32 \qquad {-4 \atop -4}$

    $\frac{7x}{3} = \frac{28}{1} \qquad 3 \times 28 = 84 =$
    $\qquad\qquad\qquad\quad 7x = 84 \quad x = 12$
    $\qquad\qquad\qquad\quad \div 7 \;\; \div 7$

    $\begin{array}{r} 28 \\ \underline{\times 28} \\ 56 \\ \underline{28} \\ 84 \end{array}$

4. $1/4x + 1/3x =$

    $\frac{1}{4}x + \frac{1}{3}x = \frac{3}{12}x + \frac{4}{12}x = \frac{7}{12}x$

5. $3/5x - 2/5x = 20$

    $\frac{3x}{5} - \frac{2x}{5} = 20$

    $\frac{x}{5} = 20$

    $x = 100$

6. Solve for z when x = 3
   $z = \dfrac{x^3}{3} - 9$

   $z = \dfrac{3^3}{3} - 9 = 6 - 9 = -3$

7. If $x/7 = 4/y$, what is the value of xy?

   $x \cdot y = 28$

Mathematics Knowledge
Algebra with Fractions - Pre-Test with Answers

1. $4/5x = 20$

   To get rid of a fraction in front of a variable, multiply by the reciprocal on both sides. Multiplying by the reciprocal cancels out the fraction in front of the variable.

   $5/4 \cdot 4/5x = 20/1 \cdot 5/4$
   $x = 100/4$
   $x = 25$

2. $5/2x - 3 = 7$

   Add 3 to both sides of the equation.
   $5/2x - 3 + 3 = 7 + 3$
   $5/2x = 10$
   Multiply by the reciprocal on both sides.
   Multiplying by the reciprocal cancels out the fraction in front of the variable.

   $2/5 \cdot 5/2 x = 10/1 \cdot 2/5$
   $x = 20/5$
   $x = 4$

3. $7/3x + 4 = 32$

   Subtract 4 from both sides.

   $7/3x + 4 - 4 = 32 - 4$
   $7/3x = 28$

   Multiply by the reciprocal on both sides.
   Multiplying by the reciprocal on both sides of the equation cancels out the fraction in front of the variable.

   $3/7 \cdot 7/3x = 28/1 \cdot 3/7$
   $x = 28/1 \cdot 3/7$
   Reduce the 28 and 7 to become 4 and 1.
   $x = 4/1 \cdot 3/1 = 12/1$
   $x = 12$

4. $1/4x + 1/3x =$

   Add the 2 fractions.
   Find a common denominator, and whatever you do to the denominator, you do to the numerator. The least common denominator is 12.
   $1/4 = 3/12$
   and $1/3 = 4/12$
   $3/12\,x + 4/12x = 7/12x$

5. $3/5x - 2/5x = 20$

   This problem has a common denominator so you can subtract.
   $1/5x = 20$

   To get rid of the fraction in front of the variable, multiply by the reciprocal.

   $5/1 \cdot 1/5\,x = 20/1 \cdot 5/1$
   $x = 100$

6. Solve for z when $x = 3$
   $z = \dfrac{x^3}{3} - 9$

   Plug 3 in for x.
   $z = \dfrac{(3)^3}{3} - 9$

   $z = \dfrac{27}{3} - 9 = 9 - 9 =$

   $z = 0$

7. If $x/7 = 4/y$, what is the value of xy?

   Cross multiply to solve.
   $x/7 = 4/y$

   $\boxed{xy = 28}$

Julie A. Hyers © 2020

Celsius, Fahrenheit, and Absolute Value Review

Changing Celsius to Fahrenheit

$F = 9/5 C + 32$

Example:
Convert 100 degrees Celsius to Fahrenheit (100° C).

$F = 9/5 C + 32$

Multiply first, then add.

$F = 9/5(100) + 32$

$F = 9/5 (100/1) + 32$

Multiply first, then add. Follow the rules of PEMDAS.

$F = 900/5 + 32 = 180 + 32 = 212°\ F$

Or it can be reduced before multiplying.
100 divided by 5 is 20.
Then it becomes,
$9/1(20/1) + 32$
$180 + 32 = 212°\ F$

100 degrees Celsius = 212 degrees Fahrenheit

Changing Fahrenheit to Celsius

$C = 5/9\ (F-32)$

Example:
Convert 212 degrees Fahrenheit to Celsius (212° F).

$C = 5/9\ (212-32)$

Subtract first, then multiply.

$C = 5/9(180)$

$C = 5/9(180/1)$

C = 900/9 = 100

You can reduce the fractions before multiplying.

C = 5/9(180/1)

180 can be divided by 9, which equals 20.

C = 5/1 (20/1) = 100° C

212 degrees Fahrenheit = 100 degrees Celsius

Absolute Value

Absolute Value tells the distance of a number from zero.

The symbol for absolute value is written this way:
$|-7| = 7$
This is also true. $|7| = 7$

The absolute value of -7 is 7 since -7 is 7 numbers away from 0.
The absolute value of 7 is also 7 since 7 is 7 numbers away from 0.

Absolute value is a number's distance from zero. It will always be a positive answer.

Absolute Value Questions with Algebra

$|x + 4| = 10$

When absolute value involves algebra, there are 2 possible solutions.

x + 4 = 10   or   x + 4 = -10

x = 6   or   x = -14

Absolute Value with Inequalities

$|-3x-6| \leq 12$

When absolute value involves algebra, there are 2 possible solutions.
This example is an inequality. Be careful with the direction of the sign.

The 2 choices are: the original equation with the absolute value sign removed or the negative answer, which needs the inequality sign flipped around. Then solve for x to find the 2 solutions. Also remember, when you divide by a negative number, you need to flip the inequality sign around as well.

$-3x - 6 \leq 12$ or $-3x - 6 \geq -12$

$-3x \leq 18$ $-3x \geq -6$

$x \geq -6$ $x \leq 2$

$-6 \leq x \leq 2$

Mathematics Knowledge
Celsius, Fahrenheit, and Absolute Value - Pre-Test

1. Convert Celsius to Fahrenheit by using the formula
   F = 9/5 C + 32, when the temperature is 20°C.

2. Convert Fahrenheit to Celsius by using the formula
   C = 5/9 (F-32), when the temperature is 95°F.

3. |-3| =

4. |5| =

5. |x+2| = 5

6. |-2x-2| ≤ 4

Julie A. Hyers © 2020

Mathematics Knowledge
Pre-Test
Celsius, Fahrenheit, and Absolute Value

1. Convert Celsius to Fahrenheit by using the formula
   F = 9/5 C + 32, when the temperature is 20°C.

   Follow order of operations (PEMDAS) in solving this problem. Multiply then add.

   F = 9/5(20/1) + 32
   It is possible to reduce when multiplying fractions in this case.
   9/5 x 20/1
   5 and 20 can be reduced since 5 goes into 20, 4 times.
   9/1 x 4/1 + 32
   (9 x 4) + 32
   36 + 32 = 68°F

2. Convert Fahrenheit to Celsius by using the formula
   C = 5/9 (F-32), when the temperature is 95°F.

   Work on the parentheses first, then multiply.

   C = 5/9(95-32)
   C = 5/9(63/1)

   In multiplying 5/9 x 63/1, you can reduce the 9 and 63. 9 goes into 63, 7 times.
   5/1 x 7/1 = 35°C

3. |-3| =

   This is an absolute value question. Absolute value is a number's distance from zero. It will always be a positive answer. The absolute value of -3 is 3 because -3 is 3 away from 0.
   |-3| = 3

4. |5| = 5

   Absolute value of 5 is 5 because the number 5 is 5 away from 0.

5. $|x+2| = 5$

When absolute value involves algebra, there are 2 possible solutions.

$x + 2 = 5$ or $x + 2 = -5$

$x = 3$ or $x = -7$

6. $|-2x-2| \leq 4$

When absolute value involves algebra, there are 2 possible solutions.
This example is an inequality. Be careful with the direction of the sign.
The 2 choices are: the original equation with the absolute value sign removed or the negative answer, which needs the inequality sign flipped around. Then solve for x to find the 2 solutions. Also remember, when you divide by a negative number, you need to flip the inequality sign around as well.

$-2x - 2 \leq 4$ or $-2x - 2 \geq -4$

$-2x \leq 6$ $-2x \geq -2$

$x \geq -3$ $x \leq 1$

$-3 \leq x \leq 1$

Math Knowledge Post-Tests and Answer Keys

1. Signed Numbers and Rounding
2. Squares and Cubes
3. Prime and Composite Numbers and Average
4. Triangles
5. Angles and Slope
6. Area, Circumference, and Perimeter
7. Volume and Surface Area
8. Real Life Applications of Perimeter, Area, Surface Area, and Volume
9. PEMDAS - Order of Operations, Scientific Notation, and Factorial
10. Exponents
11. Factoring and FOIL (Distribution) in Algebra
12. Solve for x
13. Basic Operations with Algebra
14. Algebra with Fractions
15. Celsius, Fahrenheit, and Absolute Value

Mathematics Knowledge
Signed Numbers & Rounding - Post-Test

1. $-1 + 3 =$

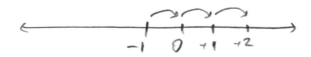

2. $-8 + -6 =$

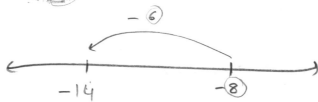

3. $-9 - 9 = -18$

4. $6 - (-1) =$ $6+1$

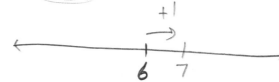

5. $-3 \times -4 = 12$

$(-)(-) = +$

6. $-2 \times 8 = -16$

7. -63 ÷ -9 = −7

8. -30 ÷ -3 =

9. Round to the nearest hundred
   45,821.93

10. Round to the nearest thousandth.
    1,943.6748

Mathematics Knowledge
Signed Numbers & Rounding - Post-Test with Answers

1. -1 + 3 = 2

2. -8 + -6 = -14

3. -9 – 9 = -18

4. 6 – (-1) = 6 + + 1 = 7

5. -3 x -4 = 12

6. -2 x 8 = -16

7. 63 ÷ -9 = -7

8. -30 ÷ -3 = 10

9. Round to the nearest hundred
   45,821.93

   45,800

10. Round to the nearest thousandth.
    1,943.6748

    1,943.675

Mathematics Knowledge
Squares and Cubes - Post-Test

1. $\sqrt{225} =$

2. $12^2 =$

3. $(-9)^2 =$

4. $\sqrt[3]{-64} =$

5. $5^3 =$ 125

6. $(-3)^3 =$ −27

7. $(-4)^2 =$ 16

8. $-(4^2) =$

9. $\sqrt{108} =$

10. $\dfrac{5\sqrt{3}}{\sqrt{2}} =$

Mathematics Knowledge
Squares and Cubes - Post-Test with Answers

1. $\sqrt{225} =$
   A square root of a number is a number that when multiplied by itself, gives the number.
   The square root of 225 is +15 or -15 because 15 x 15 = 225 or -15 x -15 – 225.

2. $12^2 =$
   12 x 12 = 144

3. $(-9)^2 =$
   ⁻9 x -9 = 81

4. $\sqrt[3]{-64} =$
   The cube root of a number is a number that when used in multiplication three times results in that number.
   The cube root of the -64 is -4 because -4 x -4 x -4 = -64.

5. $5^3 =$
   5 x 5 x 5 = 125

6. $(-3)^3 =$
   -3 x -3 x -3 = -27

7. $(-4)^2 =$
   $-4 \times -4 = 16$

8. $-(4^2) =$
   $-(4 \times 4) = -16$

9. $\sqrt{108}$

   Work to find a perfect square that goes into 108 to break it down.

   $\sqrt{36} \cdot \sqrt{3} = 6\sqrt{3}$

10. $\dfrac{5\sqrt{3}}{\sqrt{2}}$

    A radical cannot remain in the denominator. To get rid of it, multiply by $\sqrt{2}$ on the top and bottom.

    $\dfrac{5\sqrt{3}}{\sqrt{2}} \times \dfrac{\sqrt{2}}{\sqrt{2}} = \dfrac{5\sqrt{6}}{2}$

Mathematics Knowledge
Prime & Composite Numbers, Average - Post-Test

1. Which of the following is a prime number?
   A. 501
   B. 502
   C. 503
   D. 505

2. Which of the following is a composite number?
   A. 41
   B. 43
   C. 47
   D. 49 → 7·7

3. Which of the following are similar figures?
   A. cat and dog
   B. bus and trolley
   C. car and scale model of car
   D. key and lock

4. What is the mean of these numbers?
   5, 10, 10, 11, 13, 17

5. What is the median of these numbers?
   5, 10, 10, 11, 13, 17

6. What is the mode of these numbers?
   5, 10, 10, 11, 13, 17

7. What is the prime factorization of 120?

8. $\dfrac{8 \text{ yd} + 2 \text{ ft}}{2} =$

9. What is the average of 1/4 and 1/6?

Julie A. Hyers © 2020

Mathematics Knowledge
Prime & Composite Numbers, Average - Post-Test with Answers

1. Which of the following is a prime number?
   A. 501
   B. 502
   C. 503
   D. 505

   Know the divisibility rules for 2, 3 and 5.

   A number is divisible by 2 if it is even. Eliminate even numbers. The only even number that is prime is the number 2. Eliminate 502.

   A number is divisible by 3 if the sum of the digits is divisible by 3.
   501 is 5 + 0 + 1 = 6 and 6 is divisible by 3, therefore 501 is divisible by 3. Eliminate 501.

   A number is divisible by 5 is the last digit is 5 or 0. 505 ends with a 5 so it is divisible by 5. Eliminate 505.

   The remaining answer is the prime number C. 503

2. Which of the following is a composite number?
   A. 41
   B. 43
   C. 47
   D. 49

   A composite number is a number that can be divided by at least one other number aside from 1 and itself. 49 can be divided by 7, therefore it is a composite number.
   The correct answer is D. 49

3. Which of the following are similar figures?
   A. cat and dog
   B. bus and trolley
   C. car and scale model of car
   D. key and lock

   Similar figures are objects that have the same shape but may have the same size or different sizes.
   Choice C, the car and scale model of the car are similar figures.

4. What is the mean of these numbers?

    5, 10, 10, 11, 13, 17

    To find the mean (average), add up the numbers and divide by how many numbers there are.
    5 + 10 + 10 + 11 + 13 + 17 = 66
    66 ÷ 6 =
    Mean = 11

5. What is the median of these numbers?

    5, 10, 10, 11, 13, 17

    To find the median of a list of numbers, list the numbers in order from least to greatest. The middle number is the median. When there is an even amount of numbers, take the 2 middle numbers, add them up and divide by 2.

    10 + 11 = 21
    21 ÷ 2 =
    Median = 10.5

6. What is the mode of these numbers?

    5, 10, 10, 11, 13, 17

    The mode is the most common number in a list of numbers.

    In this case, the mode is 10.

7. What is the prime factorization of 120?

    To find the prime factorization of 120, divide the number up until all the factors are prime.
    120 = 12 x 10
    100 = 2 x 6 x 2 x 5
    100 = 2 x 2 x 5 x 6
    The prime factorization of 120 is 2^2 x 5 x 6

8. $\dfrac{8 \text{ yd} + 2 \text{ ft}}{2} =$

   Change the yards to feet.
   There are 3 ft in a yard.
   $\dfrac{24 \text{ ft} + 2 \text{ ft}}{2} = 26 \text{ ft}/2 = 13 \text{ ft}$

9. What is the average of 1/4 and 1/6?

   To find the average of 1/4 and 1/6, add the two fractions and divide by 2.

   If you use the least common denominator of 12, here is the solution.
   1/4 + 1/6
   3/12 + 2/12 = 5/12
   5/12 ÷ 2 = 5/12 ÷ 2/1 = 5/12 x 1/2 = 5/24

   If you use 24 as a common denominator, you end up with the same solution in the end.
   1/4 + 1/6
   6/24 + 4/24 = 10/24
   10/24 ÷ 2 = 10/24 ÷ 2/1 = 10/24 x 1/2 = 10/48 = 5/24

Math Knowledge
Triangles - Post-Test

1. What is the measure of 2 angles in an equilateral triangle?

2. If one of the equal angles of an isosceles triangle is 40°, what is the measure of the vertex angle?

3. In a right triangle, if one of the angles measures 35°, what is the measure of the third angle?

4. If a right triangle has sides that measure 3 and 4, what is the measure of the longest side?

5. Vanessa walks 10 blocks north and 24 blocks east. If a straight line is drawn from her starting point to her ending point, how many blocks is Vanessa from her starting point?

Julie A. Hyers © 2020

Mathematics Knowledge
Triangles - Post-Test with Answers

1. What is the measure of two angles in an equilateral triangle?

    An equilateral triangle has 3 equal angles. The total measure of the angles in a triangle is always 180°. An equilateral triangle has 3 equal angles (and 3 equal sides) making each angle a 60° angle.

    Two angles in an equilateral triangle are 60° + 60° = 120°.

2. If one of the equal angles of an isosceles triangle is 40°, what is the measure of the vertex angle?

    An isosceles triangle has 2 equal sides and 2 equal angles. If one of the equal angles is 40°, the other equal angle is 40°. To find the final angle, subtract the measure of the other 2 angles from 180°. 40 + 40 = 80° and 180° - 80° = 100°.

    The measure of the vertex angle is 100°.

3. In a right triangle, if one of the angles measures 35°, what is the measure of the third angle?

    In a right triangle, one angle is 90°. The other angle is 35°. To find the third angle, add up 90° + 35° = 125°, and subtract it from 180°.

    180° - 125° = 55°

    The third angle is 55°.

4. If a right triangle has sides that measure 3 and 4, what is the measure of the longest side?

    Use the Pythagorean theorem to solve this question about the sides of a right triangle.

    $c^2 = a^2 + b^2$
    $c^2 = 3^2 + 4^2 = 9 + 16 = 25$
    $c^2 = 25$

$c = \sqrt{25}$
$c = 5$

To make things easier, memorize the common right triangles.

3:4:5   5:12:13

6:8:10   10:24:26

5. Vanessa walks 10 blocks north and 24 blocks east. If a straight line is drawn from her starting point to her ending point, how many blocks is Vanessa from her starting point?

Use Pythagorean theorem to solve this problem.

The longest side is c, and the other 2 sides are a and b.

$c^2 = a^2 + b^2$
$c^2 = 10^2 + 24^2 = 100 + 576 = 676$
$c^2 = 676$
$c = \sqrt{676}$

$c = 26$

To make this easier, make sure you remember the common right triangles.

3:4:5   5:12:13

6:8:10   10:24:26

Mathematics Knowledge
Angles and Slope - Post-Test

1. What is the complement of a 71° angle?

2. What is the supplement of a 65° angle?

3. If a line has the following 2 points (1,2) and (3, -2), find the slope of the line.

4. What is the slope of a line if the equation for the line is y = 3x - 2?

5. What is the y-intercept of a line if the equation for the line is y = x + 1?

6. Find the measure of angle 7 if angle 1 = 95°. Angle 7 is _____ degrees.

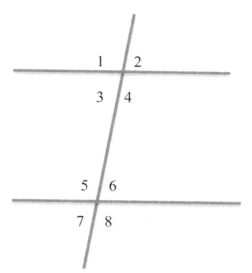

7. Which of the following is an acute angle?
   A. 25°
   B. 90°
   C. 135°
   D. 180°

8. Which of the following is an obtuse angle?
   A. 25°
   B. 90°
   C. 135°
   D. 180°

9. Which of the following is a right angle?
   A. 25°
   B. 90°
   C. 135°
   D. 180°

Mathematics Knowledge
Angles and Slope - Post-Test with Answers

1. What is the complement of a 71° angle?

   Complementary angles add up to 90°.

   90° − 71° = 19°

2. What is the supplement of a 65° angle?

   Supplementary angles add up to 180°.

   180° - 65° = 115°

3. If a line has the following 2 points (1, 2) and (3, -2), find the slope of the line.

   Point (1, 2) (x_1, y_1)

   Point (3, -2) (x_2, y_2)

   Formula for slope = $\dfrac{y_2 - y_1}{x_2 - x_1}$ $\dfrac{-2 - 2}{3 - 1}$ $\dfrac{-4}{2} = -2$

4. What is the slope of a line if the equation for the line is y = 3x - 2?

   This line is in the slope intercept form   y = mx + b.

   m is the slope, therefore the slope is 3.

   slope = 3

5. What is the y-intercept of a line if the equation for the line is y = x + 1?

   This line is in the slope-intercept form of y = mx + b

b is the y-intercept (where the line crosses the y-axis)

The y-intercept is 1.

6. Find the measure of angle 7 if angle 1 = 95°.

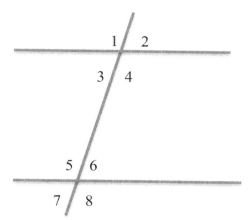

Angles that are next to each other add up to 180°.

Vertical angles, angles that are diagonally across from each other, are equal.

The angle measures of angles 1, 2, 3, and 4 are repeated in angles 5, 6, 7, and 8 respectively.

If angle 1 is 95°, angle 2 is 85°, angle 3 is 85°, angle 4 is 95,°

angle 5 is 95°, angle 6 is 85°, angle 7 is 85°, angle 8 is 95°

Angle 7 is 85°.

7. Which of the following is an acute angle?
A. 25°
B. 90°
C. 135°
D. 180°

A. 25° is the answer. An acute angle is less than 90°.

8. Which of the following is an obtuse angle?
A. 25°
B. 90°
C. 135°
D. 180°

C. 135° is the answer. An obtuse angle is bigger than 90° and less than 180°.

9. Which of the following is a right angle?
A. 25°
B. 90°
C. 135°
D. 180°

B. 90° is the answer. A right angle is equal to 90°.

Julie A. Hyers © 2020

Math Knowledge
Area, Circumference, Perimeter - Post-Test

1. What is the area of a square with a side of 5?

    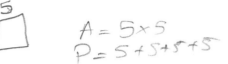

2. What is the area of a rectangle with a length of 3 and a width of 7?

    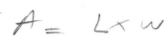

3. What is the area of a triangle with a base of 4 and a height of 8?

    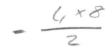

4. What is the area of a parallelogram with a base of 9 and a height of 7?

5. What is the circumference of a circle with a radius of 4?

6. What is the area of a circle with a radius of 2?

144

7. What is the area of a circle with a diameter of 10?

8. What is the area of a trapezoid with one base equal to 4, the other base equal to 10 and the height equal to 2?

9. What is the perimeter of a square with each side equal to 5?

10. What is the perimeter of a decagon when each side equals 4 cm?

11. If the perimeter of a rectangle is 100 and the length is 20, what is the area?

Math Knowledge
Area, Circumference, Perimeter - Post-Test with Answers

1. What is the area of a square with a side of 5?

    Area of a square   A = side^2 = side x side

    Side = 5
    Area = 5 x 5 =

    Area = 25

2. What is the area of a rectangle with a length of 3 and a width of 7?

    Area of a rectangle   A = length x width

    length = 3, width = 7
    Area = 3 x 7 =

    Area = 21

3. What is the area of a triangle with a base of 4 and a height of 8?

    Area of a triangle   A = 1/2 x base x height

    base = 4, height = 8
    Area = 1/2 x 4 x 8 = 2 x 8 =

    Area = 16

4. What is the area of a parallelogram with a base of 9 and a height of 7?

    Area of a parallelogram   A = bh

    Base = 9, height = 7
    A = 9 x 7 = 63

    Area = 63

5. What is the circumference of a circle with a radius of 4?

   Circumference of a circle   C = pi x diameter = πd

   Radius = 4, therefore diameter = 8
   C = 3.14 x 8 = 25.12

   Circumference = 25.12

   The ASVAB is a multiple-choice test. Before solving a question with pi in it, check the answer choices. If the answer choices are not too close in value, use 3 instead of 3.14 for π (pi). Solve quicker and choose the answer that is a little bigger than the answer you picked.

   C = 3 x 8 = 24   The estimate is 24, but the actual answer is 25.12.

6. What is the area of a circle with a radius of 2?

   Area of a circle   A = pi x radius^2 = πr^2

   Radius = 2
   A = 3.14 x 2^2 = 3.14 x 4 =

   Area = 12.56

   When you plug in 3 instead of 3.14 for π (pi), the estimate is 12. So the answer must be a bit bigger than 12 with the actual answer being 12.56.

7. What is the area of a circle with a diameter of 10?

   Area of a circle   A = pi x radius^2 = πr^2

   To find area in this case, you need to change diameter to radius.

   Radius is 1/2 the diameter. Since the diameter is 10, the radius is 5.

   Radius = 5

A = 3.14 x 5² = 3.14 x 25 =
Area = 78.5

When you plug in 3 instead of 3.14 for π (pi), you get an estimate of 3 x 25 = 75, when the actual answer is 78.5.

8. What is the area of a trapezoid with one base equal to 4, the other base equal to 10 and the height equal to 2?

   Area of a trapezoid = 1/2(base₁ + base₂)height

   Base₁ = 4, Base₂ = 10, Height = 2
   A = 1/2(4 + 10)2 = 1/2(14)2 = 7 x 2 =

   Area = 14

9. What is the perimeter of a square with each side equal to 5?

   Add up all the sides of a shape to find the perimeter.

   Perimeter = 5 + 5 + 5 + 5 = 20

10. What is the perimeter of a decagon when each side equals 4 cm?

    A decagon has 10 sides.

    Each side is 4 cm and 10 x 4 =

    Perimeter = 40 cm

11. If the perimeter of a rectangle is 100 and the length is 20, what is the area?

    The perimeter of a rectangle is 100.

    P = 2L + 2W

    Length is 20
    2L + 2W = 100
    2(20) + 2W = 100

$40 + 2W = 100$
Subtract 40 from each side.
$2W = 60$
Divide by 2.

$W = 30$

Area of a rectangle = length x width
L = 20
W = 30

Area = LW = 20 x 30 = 600

Mathematics Knowledge
Volume and Surface Area - Post-Test

1. What is the volume of a cylinder with a radius of 6 and a height of 2?

2. What is the volume of a sphere with a radius of 4?

3. What is the volume of a rectangular box with a length of 5, width of 6, and a height of 3?

4. What is the volume of a cube with a side of 4?

5. What is the surface area of a rectangular box with a length of 3, width of 4, and a height of 6?

6. What is the surface area of a cube with a side of 4?

Julie A. Hyers © 2020

Mathematics Knowledge
Volume and Surface Area - Post-Test with Answers

1. What is the volume of a cylinder with a radius of 6 and a height of 2?

   Volume of a cylinder = $\pi r^2 h$ = pi x radius^2 x height

   V = 3.14 x 6^2 x 2 = 3.14 x 36 x 2 = 3.14 x 72 =
   Volume = 226.08

   If the answer choices on the test are spaced out, you can try to use 3 instead of 3.14 for pi to get an estimate. The real answer will be a bit bigger than the estimate.
   When you use 3 instead of 3.14 for pi, the answer is 3 x 6^2 x 2 = 3 x 36 x 3 = 216.
   The estimate is 216, when the actual answer is 226.08

2. What is the volume of a sphere with a radius of 4?

   Volume of a sphere = $4/3 \pi r^3$ = 4/3 x pi x radius^3

   V = 4/3 x 3.14 x 4^3 = 4/3 x 3.14 x 64
   4/3 can be written as a decimal = 1.33
   V = 1.33 x 3.14 x 64 = 267.2768
   Volume = 267.28

   When you plug in 3 instead of 3.14 for pi, the answer is as follows,
   V = 4/3 x 3 x radius^3 = 4/3 x 3/1 x radius^3
   The two 3s cancel out, leaving the simpler formula, V = 4 x radius^3
   The simpler sphere formula is V = $4r^3$
   V = 4 x 4^3 = 4 x 64 = 256.

   The estimate is 256 when the actual answer is 267.28.
   The estimate can save time on the exam as long as the answer choices on the test are not too close in value. Make sure to pick the answer choice that is a little bigger than your estimate.

3. What is the volume of a rectangular box with a length of 5, width of 6, and a height of 3?

   Volume of a rectangular box = L x W x H

   Volume = 5 x 6 x 3 = 90

4. What is the volume of a cube with a side of 4?

   Volume of a cube = side^3

   V = s^3 = 4^3 = 4 x 4 x 4 = 64

   Volume = 64

5. What is the surface area of a rectangular box with a length of 3, width of 4, and a height of 6?

   Surface area of a rectangular box = 2(LW + LH + WH)

   L = 3, W = 4, H = 6
   SA = 2[(3 x 4) + (3 x 6) + (4 x 6)] = 2 (12 + 18 + 24) = 2(54) =

   Surface area = 108

6. What is the surface area of a cube with a side of 4?

   Surface area of a cube = 6 x side^2

   Surface Area = 6(4)^2 = 6 x 16 = 96

Mathematics Knowledge

Real Life Applications for Perimeter, Area, Surface Area, Volume - Post-Test

1. How much fence is needed to enclose a backyard with dimensions of 50 ft by 100 ft?

2. How much carpet is needed to cover the floor of a living room with dimensions of 25 ft x 20 ft?

3. How much wrapping paper is needed to cover a square box with a height of 2 inches?

4. How much wrapping paper is needed to cover a rectangular box with a length of 4 inches, a width of 7 inches and a height of 5 inches?

5. A round swimming pool has a diameter of 20 ft and a height of 6 ft. A cubic foot of water is about 7.5 gallons. How many gallons are used to fill the pool?

6. A swimming pool has dimensions of 30 ft by 40 ft by 100 ft. A cubic foot of water is about 7.5 gallons. How many gallons are used to fill the pool?

Julie A. Hyers © 2020

Mathematics Knowledge
Real Life Applications for Perimeter, Area, Surface Area, Volume
Post-Test with Answers

1. How much fence is needed to enclose a backyard with dimensions of 50 ft by 100 ft?

   A question about how much fence is needed is a perimeter question. Perimeter involves adding up all the sides of a shape. A backyard with dimensions of 50 ft by 100 ft has 4 sides: 50 ft, 50 ft, 100 ft, and 100 ft.

   For perimeter, add up the sides.
   50 ft + 50 ft + 100 ft + 100 ft =

   Perimeter = 300 ft

   Perimeter is written as yards, feet, inches, miles, etc.

2. How much carpet is needed to cover the floor of a living room with dimensions of 25 ft x 20 ft?

   A question about how much carpet is needed is an area question.

   Area of a rectangle = length x width
   25 x 20 = 500 square feet =

   Area = 500 ft^2

   Area is written as square feet (ft^2), square inches (in.^{2}), square yards (yd^2), square miles (mi^2), etc., depending on the example.

3. How much wrapping paper is needed to cover a square box with a height of 2 inches?

   A question about wrapping paper is a question on surface area.

   The surface area of a square box = 6 x side^2

   6 (2)^2 = 6(4) = 24 square feet =

   Surface area = 24 ft^2

   Surface area is written as square feet (ft^2), square inches (in.^{2}), square yards (yd^2), square miles (mi^2), etc., depending on the example.

4. How much wrapping paper is needed to cover a rectangular box with a length of 4 inches, a width of 7 inches and a height of 5 inches?

A question about wrapping paper is a question on surface area.

The surface area of a rectangular box = 2[(length x width) + (length x height) + (width x height)]
Surface area of a rectangular box = 2(LW +LH +WH)

Length = 4
Width = 7
Height = 5

2[4(7) + 4(5) + 7(5)] = 2 (28 + 20 + 35) = 2(83) = 166 square feet =

Surface area = 166 ft^2

Surface area is written as square feet (ft^2), square inches (in.^{2}), square yards (yd^2), square miles (mi^2), etc., depending on the example.

5. A round swimming pool has a diameter of 20 ft and a height of 6 ft. A cubic foot of water is about 7.5 gallons. How many gallons are used to fill the pool?

A question about how much water is in a pool is a volume question.
The volume of a round pool is a question asking about the volume of a cylinder
Volume of a cylinder = π x radius^2 x height

V = π x r^2 x h

The problem gives the diameter of the pool. Change diameter to radius to solve the problem.

If diameter is 20 ft, the radius is 10 ft.

V = π x 10^2 x 6 = π x 100 x 6 = 600π = 600 x 3.14 = 1,884 square feet = 1,884 ft^3
The pool has a volume of 1884 ft^3.

Volume is written as cubic feet (ft^3), cubic inches (in.^{3}), cubic yards (yd^3), etc., depending on the example.

To figure out how many gallons of water are in the pool, multiply the volume of 1884 ft^3 by 7.5 since there are 7.5 cubic feet of water in 1 cubic foot of water.
1884 x 7.5 = 14,130 gallons

6. A swimming pool has dimensions of 30 ft by 40 ft by 100 ft. A cubic foot of water is about 7.5 gallons. How many gallons are used to fill the pool?

A question about how much water is in a pool is a volume question.
The volume of a rectangular pool is length x width x height.

Volume = 30 x 40 x 100 = 120,000 cubic feet = 120,000 ft^3

Volume is written as cubic feet (ft^3), cubic inches ($in.^3$), cubic yards (yd^3), etc., depending on the example.

To figure out how many gallons of water are in the pool, multiply the volume of 120,000 ft^2 by 7.5 since there are 7.5 gallons of water in 1 cubic foot of water.

120,000 x 7.5 = 900,000 gallons

Mathematics Knowledge
PEMDAS, Scientific Notation, Factorial - Post-Test

1. $6 + 2 - (7-3)^2 \times 4 + 8 \div 2 =$

2. Write in scientific notation.
   3,765,431

3. Write in scientific notation.
   0.043

4. Write as a numeral.
   8.4921×10^3

5. Write as a numeral.
   6.417×10^{-3}

6. Solve.
   5!

7. Solve.
2! - 0!

8. Solve.
3 factorial

Julie A. Hyers © 2020

Mathematics Knowledge
PEMDAS, Scientific Notation, Factorial - Post-Test with Answers

1. $6 + 2 - (7-3)^2 \times 4 + 8 \div 2 =$

   Follow PEMDAS.
   Work on Parentheses first.
   Then work on Exponents.
   Solve Multiplication and Division whichever comes first in left to right order.
   Solve Addition and Subtraction whichever comes first in left to right order.

   $6 + 2 - (7-3)^2 \times 4 + 8 \div 2 =$

   $6 + 2 - (4)^2 \times 4 + 8 \div 2 =$

   $6 + 2 - 16 \times 4 + 8 \div 2 =$

   $6 + 2 - 64 + 4 =$

   $8 - 64 + 4 =$

   $-56 + 4 = -52$

2. Write in scientific notation.
   3,765,431

   3.765431×10^6

3. Write in scientific notation.
   0.043

   4.3×10^{-2}

4. Write as a numeral.
   8.4921×10^3

   8,492.1

5. Write as a numeral.
6.417 x 10^{-3}

0.006417

6. Solve.
5!

5 x 4 x 3 x 2 x 1 = 120

7. Solve.
2! - 0! = (2 x 1) – 1 =

2 – 1 = 1

8. Solve.
3 factorial

3 x 2 x 1 = 6

Mathematics Knowledge
Exponents - Post-Test

1. Evaluate.
   $3x^3 - 2y^2$
   when $x = -1$, $y = 2$

2. Evaluate.
   $3a^3 - 2b^3 - c^2$
   when $a = -2$, $b = 3$, $c = 1$

3. $\dfrac{8x^5y^2z^4}{2x^3y^4z} =$

4. $(5x^2y^4)^3 =$

5. $\dfrac{(4x^3y^5)^3}{16x^3y^4} =$

6. $(5x^3)(4x^5) =$

7. Evaluate when $x = -2$

$\dfrac{4x^4}{x^3} =$

Mathematics Knowledge
Exponents - Post-Test with Answers

1. Evaluate
   $3x^3 - 2y^2$
   when $x = -1$, $y = 2$

   $3(-1)^3 - 2(2)^2 =$
   Follow order of operations (PEMDAS)
   $3(-1) - 2(4)$
   $-3 - 8 = -11$

2. Evaluate
   $3a^3 - 2b^3 - c^2$
   when $a = -2$, $b = 3$, $c = 1$

   Follow order of operations (PEMDAS)
   $3(-2)^3 - 2(3)^3 - (1)^2$
   $3(-8) - 2(27) - (1) =$
   $-24 - 54 - 1 =$
   $-78 - 1 = -79$

3. $\dfrac{8x^5y^2z^4}{2x^3y^4z} =$

   Divide or reduce the numerals and subtract the exponents.
   $4x^2y^{-2}z^3$
   If there is a negative exponent, it must be moved down to the denominator to become positive.
   The final answer is $\dfrac{4x^2z^3}{y^2}$

4. $(5x^2y^4)^3 =$

   Raise the number 5 to the 3rd power. $5^3 = 125$
   Multiply the exponents by 3.
   $125x^6y^{12}$

5. $\dfrac{(4x^3y^5)^3}{16x^3y^4} =$

   Work on the exponent first.
   Raise 4 to the 3rd power and multiply the exponents by 3 in the numerator.

   $\dfrac{64x^9y^{15}}{16x^3y^4} =$

   Then divide or reduce the numbers and subtract the exponents.

   $4x^6y^{11}$

6. $(5x^3)(4x^5) =$

   Multiply the numbers and add the exponents.

   $20x^8$

7. Evaluate when $x = -2$

   $\dfrac{4x^4}{x^3} =$

   Simplify by dividing by x^3, which leads to $4x$.

   $\dfrac{4x^4}{x^3} = 4x = 4(-2) =$

   $x = -8$

Mathematics Knowledge
Factoring and FOIL (Distribution) in Algebra - Post-Test

1. $x^2 + 9x + 18$

2. $x^2 - 11x + 30$

3. $x^2 - 8x - 48$

4. $x^2 + 3x - 40$

5. $x^2 - 25$

6. $9x^4 - 3x^2 - 12$

7. $(x-4)(x+4)$

8. $(x+3)(x+7)$

9. (x+4)(x-2)

10. (x-5)(x-4)

11. (2x + 4)(4x – 6)

Mathematics Knowledge
Factoring and FOIL (Distribution) in Algebra - Post-Test with Answers

1. $x^2 + 9x + 18$

    Since the second sign is positive, when you factor, the two signs will be the same.
    The first sign is positive which means they will both be positive when you factor.

    What 2 numbers add to +9 and multiply to +18?
    +6 and +3
    6 + 3 = 9 and 6 x 3 = 18

    $(x + 6)(x + 3)$

2. $x^2 - 11x + 30$

    Since the second sign is positive, when you factor, the two signs will be the same.
    The first sign is negative which means they will both be negative when you factor.

    What 2 numbers add to -11 and multiply to +30?
    -6 and -5
    -6 + -5 = -11 and -6 x -5 = 30

    $(x - 6)(x - 5)$

3. $x^2 - 8x - 48$

    Since the second sign is negative, it means the 2 signs will be different when you factor.
    The first sign is negative which means the bigger number will be negative.
    What 2 numbers subtract to -8 and multiply to -48?
    -12 and +4
    -12 + 4 = -8 and -12 x 4 = -48

    $(x - 12)(x + 4)$

4. $x^2 + 3x - 40$

       Since the second sign is negative, it means the 2 signs will be different when you factor.
       The first sign is positive which means the bigger number will be positive.

       What 2 numbers subtract to +3 and multiply to -40?
       +8 and -5
       8 - 5 = 3 and 8 x -5 = -40

       (x + 8)(x - 5)

5. $x^2 - 25$

       This example is known as the difference of 2 squares since x^2 is a perfect square, 25 is a perfect square, and the operation is subtraction, also known as the difference.

       What is the square root of 25?
       +5 and -5

       (x + 5)(x - 5)

6. $9x^2 - 3x - 12$

       The second sign is negative which means that when you factor, the signs will be different.
       The first sign is negative which means the bigger number will be negative.
       The last 2 numbers need to multiply to -12

       (3x - 4)(3x + 3)

7.   (x - 4)(x + 4)

       $x^2 + 4x - 4x - 16$

       $x^2 - 16$

8. $(x + 3)(x + 7)$

   $x^2 + 7x + 3x + 21$

   $x^2 + 10x + 21$

9. $(x + 4)(x - 2)$

   $x^2 - 2x + 4x - 8$

   $x^2 + 2x - 8$

10. $(x - 5)(x - 4)$

    $x^2 - 4x - 5x + 20$

    $x^2 - 9x + 20$

11. $(2x + 4)(4x - 6)$

    $8x^2 - 12x + 16x - 24$

    $8x^2 + 4x - 24$

Mathematics Knowledge
Solve for x - Post-Test

Solve for x.

1. $3x + 4y = -6$
   $x + 2y = -4$

2. $5x - 3 = 4x + 10$

3. $6(2x + 7) = 2(7x - 7)$

4. $6(2x + 4) = 4(x + 6)$

5. $6(x + 2) = 6(x + 4)$

6.   3 – x = x – 3, then x =

7.   8x - 3 ≤ 37

8.   -9x - 1 ≥ 62

9.   -4x – 5 ≤ -53

10.   Three less than 8 times a number is 69. What is the number? Translate this problem into an algebraic equation.

Mathematics Knowledge
Solve for x - Post-Test with Answers

Solve for x.

1. $3x + 4y = -6$
   $x + 2y = -4$

   In order to solve for x, we need to work to eliminate y.

   To get rid of y, we want to make sure the numbers before y in the 2 equations are equal with opposite signs.

   We can change the second equation by multiplying the whole thing by -2.

   $-2(x+2y = -4) = -2x - 4y = 8$

   Now if we combine the two equations through addition,
   the +4y and -4y will cancel out.

   $3x + 4y = -6$
   $+ -2x - 4y = 8$
   $x = 2$

   If you need to also solve for y, plug x = 2 back into the easier equation.

   $x + 2y = -4$
   $2 + 2y = -4$
   $2y = -6$
   $y = -3$

2. $5x - 3 = 4x + 10$

   Work to get x on one side of the equal sign and the numbers on the other side of the equal sign.

   $5x - 3 = 4x + 10$
   Add 3 to both sides.
   $5x = 4x + 13$
   Then subtract 4x from both sides.
   $x = 13$

3. $6(2x + 7) = 2(7x - 7)$

    Work to get the x on one side of the equal sign and the numbers on the other side of the equal sign.

    Distribute $6(2x + 7) = 12x + 42$
    Distribute $2(7x - 7) = 14x - 14$

    $12x + 42 = 14x - 14$
    Add 14 to both sides.
    $12x + 56 = 14x$
    Subtract 12x from both sides.
    $56 = 2x$
    Divide by 2.
    $28 = x$

4. $6(2x + 4) = 4(x + 6)$

    Work to get the x on one side of the equal sign and the numbers on the other side of the equal sign.

    Distribute $6(2x + 4) = 12x + 24$
    Distribute $4(x + 6) = 4x + 24$

    $12x + 24 = 4x + 24$
    Subtract 24 from each side.
    $12x = 4x$
    Subtract 4x from each side
    $8x = 0$
    Divide each side by 8.
    $x = 0$

5. $6(x + 2) = 6(x + 4)$

    Distribute $6(x + 2) = 6x + 12$
    Distribute $6(x + 4) = 6x + 24$
    $6x + 12 = 6x + 24$
    Subtract 6x from each side.
    $12 \neq 24$

    12 cannot equal 24.

    There is no solution.

6. $3 - x = x - 3$, then $x =$

   Add 3 to both sides.
   $6 - x = x$
   Add x to both sides.
   $6 = 2x$
   Divide by 2.

   $3 = x$

7. $8x - 3 \leq 37$

   Add 3 to both sides.

   $8x \leq 40$
   Divide by 8.

   $x \leq 5$

8. $-9x - 1 \geq 62$

   Add 1 to both sides.
   $-9x \geq 63$

   Divide by -9.
   When you divide by a negative number when dealing with an inequality, flip the sign around.

   $x \leq -7$

9. $-4x - 5 \leq -53$

   Add 5 to both sides.
   $-4x \leq -48$
   Divide both sides by -4. When you divide by a negative number when dealing with an inequality, flip the sign around.

   $x \geq 12$

10. Three less than 8 times a number is 69. What is the number?

    Translate this problem into an algebraic equation.
    Let the number equal n and solve for n.

    $8n - 3 = 69$

    Add 3 to both sides.
    $8n = 72$
    Divide by 8.

    $n = 9$

Mathematics Knowledge
Basic Operations in Algebra - Post-Test

1.
$$\begin{array}{r} 5x^2 + 2x - 3 \\ +\ \underline{2x^2 - 3x + 4} \end{array}$$

2.
$$\begin{array}{r} 3x^2 + 2x + 7 \\ -\ \underline{x^2 + 4x - 3} \end{array}$$

3.
$$\frac{x^2}{x+4} \cdot \frac{x^2 + 9x + 20}{x} =$$

4.
$$\frac{x^2 - 25}{x^2} \div \frac{x+5}{x^3} =$$

Math Knowledge
Basic Operations in Algebra - Post-Test with Algebra

1.

$$\begin{array}{r} 5x^2 + 2x - 3 \\ +\ \underline{2x^2 - 3x + 4} \end{array}$$

To solve for addition with polynomials, just add up each term. Pay close attention to the signs.

$$\begin{array}{r} 5x^2 + 2x - 3 \\ +\ \underline{2x^2 - 3x + 4} \\ 7x^2 - x + 1 \end{array}$$

2.

$$\begin{array}{r} 3x^2 + 2x + 7 \\ -\ \underline{x^2 + 4x - 3} \end{array}$$

To solve for subtraction with polynomials, change subtraction to addition and change all the signs in the second expression. Place parentheses around the second expression and change all the signs inside the parentheses.

$$\begin{array}{r} 3x^2 + 2x + 7 \\ +\ \underline{(-x^2 - 4x + 3)} \\ 2x^2 - 2x + 10 \end{array}$$

3.

$$\frac{x^2}{x+4} \cdot \frac{x^2 + 9x + 20}{x} =$$

Factor the expression $x^2 + 9x + 20$ before multiplying the terms. Reduce diagonally if possible. Then multiply the numerators and the denominators straight across.

$$\frac{x^2}{x+4} \cdot \frac{(x+4)(x+5)}{x}$$

$$\frac{x^{2\ 1}}{\cancel{x+4}} \cdot \frac{\cancel{(x+4)}(x+5)}{\cancel{x}}$$

$$x(x+5)$$

4.

$$\frac{x^2 - 25}{x^2} \div \frac{x + 5}{x^3} =$$

Factor the expression $x^2 - 25$. Since this is division of polynomials in fractions, be sure to keep, change, flip. Keep the first fraction, change division to multiplication, flip the second fraction over. Reduce diagonally if possible. Multiply the numerators and the denominators straight across.

$$\frac{(x + 5)(x - 5)}{x^2} \cdot \frac{x^3}{x + 5}$$

$$\frac{\cancel{(x + 5)}(x - 5)}{x^2} \cdot \frac{x^{3\,1}}{\cancel{x + 5}}$$

$$x(x - 5)$$

Mathematics Knowledge
Algebra with Fractions - Post-Test

1. $2/3x = 20$

2. $3/4x - 2 = 7$

3. $8/5x + 2 = 34$

4. $1/5x + 1/2x =$

5. $4/6x - 1/6x = 30$

6. Solve for z when x = 4

    $z = \dfrac{x^2 - 4}{2}$

7. If $x/5 = 7/y$, what is the value of xy?

Julie A. Hyers © 2020

Mathematics Knowledge
Algebra with Fractions - Post-Test with Answers

1. $2/3x = 20$

   To get rid of a fraction in front of a variable, multiply by the reciprocal on both sides. Multiplying by the reciprocal on both sides of the equation cancels out the fraction in front of the variable.

   $3/2 \cdot 2/3x = 20/1 \cdot 3/2$
   $x = 60/2$
   $x = 30$

2. $3/4x - 2 = 7$

   Add 2 to both sides of the equation.
   $3/4x - 2 + 2 = 7 + 2$
   $3/4x = 9$

   Multiply by the reciprocal on both sides. Multiplying by the reciprocal on both sides of the equation cancels out the fraction in front of the variable.

   $4/3 \cdot 3/4x = 9/1 \cdot 4/3$
   $x = 36/3$
   $x = 12$

3. $8/5x + 2 = 34$

   Subtract 2 from both sides.

   $8/5x + 2 - 2 = 34 - 2$
   $8/5x = 32$

   Multiply by the reciprocal on both sides. Multiplying by the reciprocal on both sides of the equation cancels out the fraction in front of the variable.

   $5/8 \cdot 8/5x = 32/1 \cdot 5/8$
   $x = 32/1 \cdot 5/8$
   Reduce the 32 and 8 to become 4 and 1.
   $x = 4/1 \cdot 5/1 = 20/1$
   $x = 20$

4. $1/5x + 1/2x =$

Add the 2 fractions.
Find a common denominator, and whatever you do to the denominator, you do to the numerator. The least common denominator is 10.
$1/5 = 2/10$
and $1/2 = 5/10$

$2/10x + 5/10x = 7/10\ x$

5. $4/6x - 1/6x = 30$

This problem has a common denominator so you can subtract.
$3/6x = 30$

To get rid of the fraction in front of the variable, multiply by the reciprocal.

$6/3 \bullet 3/6x = 30/1 \bullet 6/3$
$x = 180/3$
$x = 60$

6.   Solve for z when $x = 4$
$$z = \frac{x^2}{2} - 4$$

Plug 4 in for x.

$$z = \frac{(4)^2}{2} - 4$$

$$z = \frac{16}{2} - 4 = 8 - 4 =$$

$z = 4$

7.   If $x/5 = 7/y$, what is the value of xy?

Cross multiply to solve.

$x/5 = 7/y$

$xy = 35$

Mathematics Knowledge
Celsius, Fahrenheit, and Absolute Value - Post-Test

1. Convert Celsius to Fahrenheit by using the formula
   F = 9/5 C + 32 when the temperature is 40°C.

2. Convert Fahrenheit to Celsius by using the formula
   C = 5/9 (F-32) when the temperature is 68°F.

3. |-7| =

4. |4| =

5. |x-3| = 3

6. |-3x-6| ≥ 18

Mathematics Knowledge
Celsius, Fahrenheit, and Absolute Value - Post-Test with Answers

1. Convert Celsius to Fahrenheit by using the formula
   F = 9/5 C + 32 when the temperature is 40°C.

   Follow order of operations (PEMDAS) in solving this problem.
   Multiply then add.

   F = 9/5(40/1) + 32
   It is possible to reduce when multiplying fractions in this case
   9/5 x 40/1   5 and 40 can be reduced since 5 goes into 40, 8 times
   9/1 x 8/1 + 32
   9 x 8 + 32
   72 + 32 = 104°F

2. Convert Fahrenheit to Celsius by using the formula
    C = 5/9 (F-32) when the temperature is 68°F.

   Work on the parentheses first, then multiply.

   C = 5/9(68-32)
   C = 5/9(36/1)
   In multiplying 5/9 x 36/1, you can reduce the 9 and 36. 9 goes into 36, 4 times.
   5/1 x 4/1 = 20°C

3. $|-7| =$

   This is an absolute value question. Absolute value is a number's distance from zero. It will always be a positive answer.
   The absolute value of -7 is 7 because -7 is 7 away from 0.
   $|-7| = 7$

4. $|4| =$

   Absolute value of 4 is 4 because the number 4 is 4 away from 0.

   $|4| = 4$

5. |x-3| = 3

   When absolute value involves algebra, there are 2 possible solutions.

   x - 3 = 3   or   x - 3 = -3

   x = 6   or   x = 0

6. |-3x-6| ≥ 18

   When absolute value involves algebra, there are 2 possible solutions.
   This example is an inequality. Be careful with the direction of the sign. The 2 choices are: the original equation with the absolute value sign removed or the negative answer, which needs the inequality sign flipped around. Then solve for x to find the 2 solutions. Also remember, when you divide by a negative number, you need to flip the inequality sign around as well.

   -3x - 6 ≥ 18   or   3x - 6 ≤ -18
   -3x ≥ 24   -3x ≤ -12
    x ≤ -8   x ≥ 4

               x ≤ -8   or   x ≥ 4

Printed in Great Britain
by Amazon